The Artisan and the European Town, 1500-1900

Historical Urban Studies

Series editors: Richard Rodger and Jean-Luc Pinol

Titles in this series include:

Capital Cities and their Hinterlands in Early Modern Europe
edited by Peter Clark and Bernard Lepetit

*Power, Profit and Urban Land: Landownership in Medieval and
Early Modern Northern European Towns*
edited by Finn-Einar Eliassen and Geir Atle Ersland

*Water and European Cities from the Middle Ages to the
Nineteenth Century*
edited by Jean-Luc Pinol and Dennis Menjeot

The Built Form of Colonial Cities
Manuel Texeira

The Artisan and the European Town, 1500–1900

Edited by

GEOFFREY CROSSICK

© Geoffrey Crossick and the contributors, 1997

All rights reserved. No part of this publication may be reproduced, stored in a retrieval system, or transmitted in any form or by any means, electronic, mechanical, photocopying, recording, or otherwise without the prior permission of the publisher.

Published by
SCOLAR PRESS
Gower House
Croft Road
Aldershot
Hants GU11 3HR
England

Ashgate Publishing Company
Old Post Road
Brookfield
Vermont 05036-9704
USA

British Library Cataloguing in Publication Data

The Artisan and the European Town, 1500-1900.
 (Historical Urban Studies)
 1. Artisans — Europe — History. 2. Cities and towns — Europe — History.
 3. Europe — History — 1492.
 I. Crossick, Geoffrey.
 940.2'091732

ISBN 1-85928-232-6

Library of Congress Cataloging-in-Publication Data

The artisan and the European town, 1500-1900 / edited by Geoffrey Crossick.
     p.   cm.
  Historical Urban Studies)
  Includes index.
  ISBN 1-85928-232-6 (acid free paper)
  1. Artisans — Europe — History. 2. Cities and towns — Europe — History.
  I. Crossick, Geoffrey. II. Series.
 HD2346.E9A77   1997
 331.7'96—dc21   96-47302
                                                                      CIP

ISBN 1 85928 232 6

This book is printed on acid-free paper

Typeset in Sabon by Express Typesetters Ltd, Farnham, Surrey
and printed in Great Britain by The Ipswich Book Company Ltd, Suffolk

Contents

List of tables		ix
List of figures		x
Notes on contributors		xi
Preface		xv
1	Past masters: in search of the artisan in European history Geoffrey Crossick	1
2	Artisans and urban politics in seventeenth-century Germany Christopher R. Friedrichs	41
3	Cultural analysis and early modern artisans James R. Farr	56
4	'Broken all in pieces': artisans and the regulation of workmanship in early modern London Michael Berlin	75
5	The aristocratic *hôtel* and its artisans in eighteenth-century Paris: the market ruled by court society Natacha Coquery	92
6	Craftsmen and revolution in Bordeaux Josette Pontet	116
7	Craftsmen in the political and symbolic order: the case of eighteenth-century Malmö Lars Edgren	131
8	Women and the craft guilds in eighteenth-century Nantes Elizabeth Musgrave	151
9	Worlds of mobility: migration patterns of Viennese artisans in the eighteenth century Josef Ehmer	172
10	Artisans in Hungarian towns on the eve of industrialization Vera Bácskai	200
11	Urban renovation and changes in artisans' activities: the Parisian *fabrique* in the Arts et Métiers quarter during the Second Empire Florence Bourillon	218

12 Artisans and the labour market in Dutch provincial
    capitals around 1900   239
       Pim Kooij

Index   257

Historical Urban Studies
General Editors' Preface

Density and proximity of buildings and people are two of the defining characteristics of the urban dimension. It is these which identify a place as uniquely urban, though the threshold for such pressure points varies from place to place. What is considered an important cluster in one context - a few hundred inhabitants or buildings on the margins of Europe - may not be considered as urban elsewhere. A third defining characteristic is functionality - the commercial or strategic position of a town or city which conveys an advantage. Over time, these functional advantages may diminish, or the balance of advantage may change within a hierarchy of towns. To understand how the relative importance of towns shifts over time and space is to grasp a set of relationships which is fundamental to the study of urban history.

Towns and cities are products of history, yet have themselves helped to shape history. As the proportion of urban dwellers has increased, so the urban dimension has proved a legitimate unit of analysis through which to understand the spectrum of human experience and to explore the cumulative memory of past generations. Though obscured by layers of economic, social and political change, the study of the urban milieu provides insights into the functioning of human relationships and, if urban historians themselves are not directly concerned with current policy issues, few contemporary concerns can be understood without reference to the historical development of towns and cities.

This longer historical perspective is essential to an understanding of social processes. Crime, housing conditions and property values, health and education, discrimination and deviance, and the formulation of regulations and social policies to deal with them were, and remain, amongst the perennial preoccupations of towns and cities - no historical period has a monopoly of these concerns. They recur in successive generations, albeit in varying mixtures and strengths; the details may differ but the central forces of class, power and authority in the city remain. If this was the case for different periods, so it was for different geographical entities and cultures. Both scientific knowledge and technical information were available across Europe and showed little respect for frontiers. Yet despite common concerns and access to broadly similar knowledge, different solutions to urban problems were proposed and adopted by towns and cities in different parts of Europe. This comparative dimension informs urban historians as to which were systemic factors and which were of a purely local nature: general and particular forces can be distinguished.

These analytical frameworks, considered in a comparative context, inform the books in this series.

Jean-Luc Pinol
Richard Rodger

Université des sciences humaines de Strasbourg
University of Leicester

List of tables

5.1	Distribution by *quartier* and trade sector of suppliers of five aristocratic families	97
5.2	Distribution by trade sector of the suppliers of five aristocratic families	98
7.1	Seating of men in St Peter's Church, Malmö, in 1728 and 1771	143
7.2	Social distinctions in funeral practices in St Peter's parish, Malmö, 1771	145
9.1	Geographic origins of Viennese guild masters, 1742	180
9.2	Percentage of Viennese guild masters born in Vienna, 1742	182
9.3	Geographic origins of Viennese apprentices	185
9.4	Geographic origins of Viennese journeymen	187
10.1	Distribution of masters in Pest and Debrecen according to income tax paid	207
10.2	Distribution of masters in Esztergom, Gyôr and Szeged according to income tax paid in the 1840s	208
10.3	Distribution of merchants in Pest, Esztergom, Gyôr and Debrecen according to income tax paid in the 1840s	208
11.1	Tables and categories of the *patente* in the Arts et Métiers quarter, 1852-79	225
12.1	Mechanized and craft firms in Groningen	245
12.2	Composition of the Groningen group of artisans (masters and men)	248
12.3	Occupations of Groningen immigrants 1900-1910	250
12.4	Occupations of male members of Groningen 1880 birth cohort at age 30	252
12.5	Occupational structure of Groningen heads of households (1910), immigrants (1900-10) and birth cohort members (1910)	253

List of figures

5.1	The tradesmen of Princess Kinsky, rue Saint-Dominique (1773-94), all trades	96
5.2	Distribution of the Coigny family's tradesmen, porte Saint-Honoré (1770-75)	99
5.3	The tradesmen of Princess Kinsky, rue Saint-Dominique (1773-94), housing trades	102
5.4	The tradesmen of the La Trémoille family, rue Saint-Dominique (1781-92), horses sector	106
11.1	Major works in Paris during the Second Empire	219
11.2	Major works in the Arts et Métiers quarter	222

Contributors

Vera Bácskai is Professor of History at the Eötvös Lóránd University of Budapest. After her initial researches on medieval and early modern Hungarian market towns, she has published essays on the economic and social history of Hungarian towns in the eighteenth and nineteenth centuries, the shaping of urban networks, and the rise of the middle class in Central Europe. Her books, besides those published in Hungarian, include *Towns and Urban Society in Early Nineteenth-Century Hungary* (1989) and *Bürgertum und bürgerliche Entwicklung in Mittel- und Osteuropa* (editor, 1986).

Michael Berlin teaches history at Middlesex University and is a research officer at the Centre for Metropolitan History at the University of London. He is the author (with Robert Iliffe, Derek Keene and David Mitchell) of *The Growth of the Skilled Workforce in Early Modern London* (forthcoming).

Florence Bourillon is Maître de Conférences at the University of Paris XII - Val de Marne. Her interests focus on urban transformation in nineteenth-century France, in particular the reconstruction carried out in Paris under the Second Empire, and the formation and development of suburbs. Her thesis was concerned with the Arts et Métiers quarter in Paris, and in addition to contributions to various collections of essays she has written *Les villes en France au XIXe siècle* (1992).

Natacha Coquery is Maître de Conférences at the University of Tours. Her doctoral thesis, 'De l'hôtel aristocratique aux ministères: habitat, mouvement, espace à Paris au XVIIIe siècle' (1995), was concerned with aristocratic society in eighteenth-century Paris, with particular emphasis on questions of economic and urban power. She has published various articles on the emergence of administrative quarters in Paris, the aristocratic consumer market, the beginnings of advertising, and the aristocratic *hôtel* and the urban patrimony.

Geoffrey Crossick is Professor of History and Dean of the Graduate School at the University of Essex. After his initial research on English working-class history, he has published widely on European social and urban history, most notably on the petite bourgeoisie and on comparative history. In addition to articles in many journals and edited collections, his books include *The Lower Middle Class in Britain 1870-1914* (editor, 1977), *An Artisan Elite in Victorian Society: Kentish London 1840-1880* (1978), *Shopkeepers and Master Artisans in Nineteenth-Century Europe* (edited with Heinz-Gerhard Haupt, 1984) and, most recently, with Heinz-Gerhard Haupt, *The Petite Bourgeoisie in Europe 1780-1914: Enterprise, Family and Independence* (1995).

Lars Edgren is Lecturer in History at the University of Lund. His research focuses on the history of the working class in nineteenth-century Sweden, with a particular interest in artisans and the transformation of guild organization. His articles include 'Crafts in transformation?: Masters, journeymen, and apprentices in a Swedish town, 1800-1850', *Continuity and Change*, 1, 1986. His principal work is *Lärling - gesäll - mästare. Hantverk och hantverkare i Malmö 1750-1847* (1987), (Apprentice - Journeyman - Master. Crafts and Craftsmen in Malmö 1750-1847, with English summary).

Josef Ehmer is Professor of Modern History at the University of Salzburg. After his initial research on the social and economic history of Vienna, he has published widely on the history of family and demography, and on labour history. His major interest is the comparative history of Central and Western Europe. He is the author of *Familienstruktur und Arbeitsorganisation im frühindustriellen Wien* (Family structures and labour relations in early industrial Vienna) (1980); *Sozialgeschichte des Alters* (Social History of Old Age) (1990); *Heiratsverhalten, Sozialstruktur, ökonomischer Wandel. England und Mitteleuropa in der Formationsperiode des Kapitalismus* (Marriage patterns, social structure and economic change. England and Central Europe in the period of capitalist formation) (1991); and *Soziale Traditionen in Zeiten des Wandels. Arbeiter und Handwerker im 19. Jahrhundert* (Social traditions in times of change. Workers and artisans in the nineteenth century) (1994).

James R. Farr is Professor of History at Purdue University and Co-Editor of *French Historical Studies*. He has published on the history of work and on gender and legal history, concentrating on the early modern period. His books include *Hands of Honor: Artisans and their World in Dijon, 1550-1650* (1988) and *Authority and Sexuality in Early Modern Burgundy (1550-1730)* (1995). He is currently completing a general work entitled *Butchers, Bakers and Candlestickmakers: Artisans in Europe from the Black Death to the Industrial Era* to be published by Cambridge University Press, and is engaged in a project on legal cultural history entitled *The Giroux Affair: Murder, Passion and the Law in Seventeenth-Century France*.

Christopher R. Friedrichs is Professor of History at the University of British Columbia in Vancouver. Much of his research concerns the social and political history of German cities in the early modern period. In addition to numerous articles, he has published two books: *Urban*

Society in an Age of War: Nördlingen 1580-1720 (1979) and *The Early Modern City, 1450-1750* (1995).

Pim Kooij is Professor of Urban and Regional History at the University of Groningen. He has published various books and articles on the theory of urban history. His empirical research has been primarily concerned with the northern Netherlands, in the joint project Integral History of the universities of Groningen, Utrecht, Yaroslavl and Tambov (in Russia). This project will publish *Where the Twain Meet. Demographic Development in Russia and the Netherlands*. He contributed the chapter on the Netherlands in R. Rodger, ed., *European Urban History* (1993), and edited *Regional Capitals: Past, Present, Prospects* (1994).

Elizabeth Musgrave is Lecturer in History at Nene College, Northampton. After completion of a doctoral thesis on the building industries in early modern Brittany, she has undertaken further research on industries and artisans in seventeenth- and eighteenth-century France, with particular reference to women's work in the urban economy. Recent articles have been published in *Histoire sociale/Social History, French History*, and *Construction History*.

Josette Pontet is Professor of Early Modern History and Director of the Centre d'Études des Espaces Urbains (Histoire) at the University of Bordeaux III. Her research has been concerned with urban, social and demographic history. After publication of her thesis, *Bayonne, un destin de ville moyenne à l'époque moderne* (1990), she edited *Histoire de Bayonne* (1991) and contributed to various volumes including *La France et la Mer* (1993) and *Ville et Port, XVIIIe-XXe siècles* (1994).

Preface

The essays in this collection focus on the artisanal presence and the artisanal experience in Europe between the beginning of the early modern period and the end of the nineteenth century. They range widely, from the masters who served the most demanding of aristocratic clients to the journeymen who travelled round Central Europe in search of work and opportunities, from ritual representations of the social order to the fate of artisans and their guilds in a major port city during the French Revolution, from artisans' political activities to the presence and difficulties of women artisans, and much more. The majority of the essays were presented in earlier versions at a session which I organized at the Conference of the European Association of Urban Historians, held in Strasbourg in 1994. All have been extensively rewritten for publication in this book, and some new essays have been added.

The majority of the contributors to this book were present in Strasbourg, and at the opening plenary session of the conference heard a typically provocative, witty and innovative lecture by Bernard Lepetit. As this book was in its final stages of preparation, news arrived of his tragic death in an accident in Paris. French urban history in particular, and European urban history as a whole, has lost one of its leading figures, a historian of imagination and insight who had so much more to contribute. This book is dedicated to his memory.

Geoffrey Crossick
Colchester, 1996

In memory of Bernard Lepetit

CHAPTER ONE

Past masters: in search of the artisan in European history

Geoffrey Crossick

Visions of the past

Artisans were rather like villages. In the modern European imagination they came to represent a world in which harmony and community ruled, in contrast with the inadequacies of urban industrial society. It is taken for granted that in an unspecified past things were better. At the end of the nineteenth century, when urban consumers used images of a harmonious and natural countryside to reassure themselves in the face of the anxieties induced by urban society, the postcards which they bought in such profusion were often of traditional village craftsmen and craftswomen. The artisan and the village came to occupy similar spaces in the urban imagination as inversions of the menace of modernity. These idealizations of the past pose problems for the historian of the artisan, for the implications of the term go beyond the merely occupational. The implicit meanings of artisanship include expectations about the nature of work and workplace relations, place within the urban social order, the role of family, and much else. Artisans, like villages, evoked interest at the end of the nineteenth century because they were in decline, and because their past seemed to represent an alternative to the harsher face of modernity. The meanings of artisanship were thus embedded in a particular reading of the past,[1] and historical study of artisans must take account of the myths which successive generations have woven around them and which they indeed wove around themselves - myths that were about the past.

One dimension of this renewed later nineteenth-century interest in the artisanal past was the publishing of craft histories. Social conservatives who saw craft masters as a bulwark with which to defend property, and masters themselves whose pride in their craft's past grew as its present became less comfortable, combined to reconstruct artisanal traditions. The scholarly Albert Babeau and a host of lesser mythologizers wrote widely cited artisanal histories which evoked the artisanal workshop and household of former times as the basis for social, economic, and indeed aesthetic order. The individualism unleashed by the modern world was contrasted with these ordered pasts.

> If this [corporate] system was contrary to civil and economic freedom, it favoured security for the worker. In the corporation he found a craft family which raised him as an apprentice and then supported him as a master; he also found his civic importance there, something which since he became isolated he has never been able to enjoy.[2]

Babeau wove together the themes of patriarchal household, continuity of place, personal relations with customers, the simplicity of the honest trader, care for apprentices, workplace discipline, and the wife's lack of vanity and involvement in the business, to present what he acknowledged was a distinctive vision. 'For those who are in favour of the present', he wrote, 'the past is barbarism; for others it is the ideal', and his writings were cited by those for whom the artisanal ideal represented a social alternative.[3] Histories of individual crafts proliferated. François Husson wrote histories of various trades, and the period since 1789 was significantly given a mere 22 pages out of the 225 in his book on joiners.[4]

Artisanal movements were themselves active in constructing identities through history. The master bakers of Vienna reorganized their archive to make it worthy of the corporation, and published their history, as did the city's carpenters' guild. Festival processions were revived with traditional costumes and old guild banners.[5] In France the dual impulse of the regionalist movement seeking to establish the traditional *pays* and the revival of historic ideas of craft production produced a flurry of local craft museums, while by the 1890s Swedish craft organizations were opening museums and commissioning their craft histories.[6] In the later nineteenth century almost every issue of the Swedish masters' journal, *Handtverks- och Industri- Tidning*, carried excerpts from old guild archives.[7] The motives behind these activities varied, but all drew upon a belief that artisanship offered a social as well as an economic vision to set against contemporary society, a vision which museums and histories might affirm and protect. There was also an aesthetic dimension to this artisanal alternative. In Austria, Camillo Sitte used his headship of the State Trade Schools in Salzburg, and then Vienna, to articulate a quasi-medieval version of artisanal culture, advocating the integrating power of artisanal values to overcome the fragmentation of modern life.[8]

Social catholic and Le Playist conservative discourse between the 1890s and 1914 provides a striking example of the way the meaning of artisanship came to be rooted in a particular reading of the past.[9] For the Comte de Boissieu, the normal, prosperous, family workshop ... teaches its members the value of the family home, maintains an intimacy between husband and wife and within the family which is always endangered by factory life, and ensures the recruitment of an elite of workers, endowed with the happy qualities of initiative and independence.[10]

Another writer cast the vision explicitly in the past:

> In former times, the *patron* was a craftsman who loved his occupation just as an artist loved his art. He was less driven by the demands of frantic competition, and had the time to attend to his apprentices. And the child, sharing his master's life, sleeping beneath his roof and sharing his table, made his way gently but surely towards becoming a master himself. From morning until night he benefited from the lessons of his master who, immersed above all in love of his craft and of his guild, saw in his apprentice a pupil and the continuation of his work [*œuvre*].[11]

These writers were attached to a distinctive social catholic perspective, but their vision reflected broader assumptions about the character of artisanal life in the past. The historical ideal of the artisan has proved less capable of inspiring social alternatives since the First World War - the enlargement of the *classes moyennes* to embrace white-collar employees shows the search for an alternative conception of the middle[12] - but artisanal organizations and the need for artisanal votes kept the vision in public view, to be renewed by various right-wing regimes, none more so than in Vichy France. Pétain proclaimed his desire to return the artisans, and all they stood for, to their traditional place in French society. 'Class struggle is impossible in the artisanal workshop', he proclaimed in 1942 during his May Day speech on the artisanate, endorsing the historic vision of workshop production, even if this Pétainist rhetoric was rarely matched by policy.[13]

There had always been a better artisanal past, it seems, whether in the minds of craftsmen themselves or of those seeking to sketch a better future from an idealized artisanate. The success of Reformation theology in early sixteenth-century Augsburg rested, according to Roper, on its appeal to urban craftsmen, as it 'imagined a civic haven in which the household would be restored to its mythical, ordered past'.[14] In his *Tableau de Paris*, published in the early 1780s, Mercier bemoaned the way the old family-like ties in artisanal production had been replaced by money and market alone. The links in the chain had gone, journeymen could move at will, and the old world of harmony and order was no more.[15] The eighteenth-century Parisian glazier, Jacques-Louis Ménétra, was convinced that in the old days social relations in the trade were more harmonious, and Roche sees this belief as one of the governing notions transmitted through the journeymen's *tour de France*.[16] Swedish master artisans felt the same at the end of the nineteenth century, complaining that the freedom of trade established in mid-century had destroyed harmonious relations in workshop production. 'Instead of being, as in the past, members of the same family, employers and workers now face each other as distrustful and often bitter enemies.'[17] We return to the late nineteenth century, as if the myth of artisanal decline might be found at almost any time in the European past. It is easy for historians to mythologize the period before their own study begins, and this tendency is exacerbated for the historian of artisans by generations of writing about the fate of the crafts.[18] Guild abolition and industrialization may have made decline the obvious template upon which to sketch artisanal history in the eighteenth and nineteenth centuries, but it applies to earlier periods as well. Those in search of the artisans in European history must be wary of visions of the past.

The meanings of artisanship

'I was born on 13 July 1738 a native of this great city. My father belonged to the class usually called artisans. His profession was that of glazier.'[19] Thus Jacques-Louis Ménétra opened the journal of his life. If artisanship was inseparable from its memories and its vision of the past – as it was even for a craftsman like Ménétra, frustrated by the restrictions imposed by his guild – then we must ask what it meant to be an artisan. At the heart, according to Zarca's socioeconomic definition, lay a technical division of labour in production that left distinguishable crafts to carry out the making of distinct products or their parts. A craft was a body of producers tied together by a set of techniques and knowledge which could be acquired only through the practice of the occupation itself over time. The artisan or craftsman need not make the whole product, but the division of labour needed to rest on distinguishable crafts. The artisan could produce directly for the market as a small entrepreneur, to order from individual consumers, or on order from merchants, or exclusively for a single merchant, or sell not the product but his or her labour to a master artisan or putting-out merchant. In the last of these the artisan was a wage-earner, in the other cases some kind of master.[20] Such a definition can only set general parameters, for artisans' situations varied greatly during the four centuries with which this collection of essays is concerned: in terms of the organization of the craft and labour process, labour force composition, household structure, the nature of the contracts and process of remuneration, the gender composition of crafts, their juridical basis, their significance to the economy, and much more.

Artisanship came to refer to distinct but overlapping phenomena. The actual work which artisans did was not necessarily, and certainly not always, the point around which they were constituted, but artisans would exist in neither our sources nor our imagination had they not been performing an economic activity within production or the provision of services. Artisanship came to mean much more than work, as Farr argues in his essay (Chapter Three), and it carried meanings that were often detached from work, but without its economic functions it would not appear upon our agenda. What were these overlapping phenomena? First, there was artisanship as occupation: the job carried out by an individual to earn his or her living, and its characteristics. Second, there was artisanship as social position: the place within a social order that accompanied the designation 'artisan'. Third, there was artisanship as identity: the sense of personal and family identity associated with being an artisan and the meaning for the sense of self. All three dimensions are relevant to our concern for the urban artisan, but it might be argued that the trilogy is in part the product of the kind of sources through which artisans have been studied. Occupational title came to be used to denote position in the urban administrative, legal or social order, and historians have inevitably focused on artisanal occupations and attached to them a degree of social significance that can be matched in no other sector of society. If one was an artisan – certainly if one was a male artisan – it has come to be assumed that artisanship was at the heart of one's social being and personal identity.

The different sources which have been used by those studying artisans have all served to encourage historians in that direction. Memoirs written by artisans such as Ménétra and Perdiguier give a privileged place to the primacy of their trade,[21] as does the discourse of protest movements, whether the struggles of journeymen to defend their craft position in the eighteenth and early nineteenth centuries, or the masters' defence movements which developed towards the end of the century. Indeed, the French philosopher and historian Jacques Rancière has argued that the discourse of French artisan radicals in the early nineteenth century should not be read as an expression of their artisanal identity. They were generally from the most debased crafts, such as tailors and shoemakers, and their conception of work derived not from their own experience, for they had little work commitment of their own, but from their search for identity through constructing the myth of the proud artisan under challenge. For Rancière, 'a strong militant identity among workers in a craft seems to imply a weak collective professional identity and vice versa'.[22]

Whether or not we accept Rancière's conclusion, his insight alerts us to the way our sources encourage the view that occupational identity was the core of artisans' existence. The introspection and cohesion of artisanal life have similarly been too readily extrapolated from the image which urban guilds presented of themselves. Instead of seeing craft guilds as an expression of occupational structures and artisanal identities, studies of late medieval York and seventeenth- and early eighteenth-century Turin both invite us to understand them as institutions constructed for political and administrative purposes, carving up the urban workforce in ways that inadequately capture either the diversity of attachment to craft cultures or the multiple ways in which individual and family incomes were made. The stress on individual crafts and guilds often came not from the daily life of artisans, but from the needs of urban government, of middling groups seeking to protect their interests, and of those recording the urban population.[23] On the other hand, in the case of seventeenth-century Germany, as explored in Friedrichs's essay (Chapter Two), guild and craft solidarities did indeed reinforce each other. Administrative imperatives could alternatively understate the artisanal presence. In seventeenth- and eighteenth-century Sweden the rural population was recorded as peasant, crofter or cottager, even where individuals were primarily artisans, because the state sought to describe not people's occupations but their status for land tax assessment. The apparent absence of artisans from the Swedish countryside was thus partly an illusion.[24]

From the middle ages, occupational labels became the increasingly normal means to distinguish people, implicitly concealing the variety of ways in which individuals earned their living as well as their changes of trade over time.[25] The occupational tables of nineteenth-century censuses intensified the practice of viewing the social structure through individual occupational title, thus strengthening the emphasis on craft title and also producing awkward and misleading attempts to identify masters and other forms of independent artisans.[26] Even without its hesitant attempts to identify small enterprise, the census offered a distorted picture of nineteenth-century urban society as one in which each individual was classified according to occupation, which was thus privileged as the designator of personal identity, and in which each adult was classified according to a single occupation. This ignored the multiple occupations and activities, the instability of occupation over time, and the diversity of family activities, which all undermined the powerful significance of occupational title for personal identity implicit in the census. Future research on artisans needs to explore the multiplicity of links which defined individuals' circumstances and identity. Cerutti's study of Turin shows that strategies for family survival and for the development of family resources required a wide spread of occupations and occupational contacts rather than the risky strategy of turning in on a single trade. This was emphasized specifically in Turin by the need to balance sources of influence between the municipality and the ducal court.[27] Relationships of family, neighbourhood, church and patronage may all have been far more significant for individual identity than those of occupation alone. Indeed, Gribaudi and Blum's preliminary picture of the nineteenth-century French social landscape suggests that conventional groupings of occupations bear very limited relation to the image that emerges from the actual social linkages between occupations through marriages or through comparing the occupations of fathers and children.[28]

None of this should lead one to dismiss artisanal identity as a mere construction. There were artisans practising their crafts, and they constituted a significant part of the production process throughout the period with which this book is concerned. However, the historic ideal of the artisan and the sources on which the historian depends conspire to reinforce the belief that their craft was the fundamental experience for these artisans. Four features were of particular significance within this classic artisanal ideal: artisanship as a lifetime project; the artisanal household as fundamental to craft production; the craft as the basis of personal and social identity; and the masculine character of artisanship.

Artisanship as a lifetime project was fundamental to the pre-industrial idea of the artisan, and expectations created then persisted long after the structures which had sustained them had disappeared. In the classic *cursus honorum* a young man began as an apprentice, then spent a period as a journeyman, during which several years would be spent travelling, and finally set up as a master on his own account, having satisfied the jurors of the corporation of his skill by producing a piece of work, literally a masterpiece. The expectation that becoming a master coincided with setting up both household and enterprise remained the norm in early modern Europe. Apprentices would live in their master's household, as would many journeymen, though this practice was under strain in much of Europe during the eighteenth century, even in Germany where conventional artisanal household structures were more resilient.[29] In many countries journeymen would be expected to spend a period of time travelling around the country, or even around Europe. The *tour de France* may not have been obligatory, but its established routes and attendant rituals were maintained by journeymen's associations known as *compagnonnages*. The German *Wanderjahre*, on the other hand, was compulsory in guild and sometimes state regulations, yet lacked the ritualized structure found across the Rhine. Its operation in Austria is closely examined in Ehmer's essay (Chapter Nine). For many journeymen travel was a way to obtain experience of the trade before setting up in one's town of origin, or a means to find openings elsewhere. Yet journeymen were also young: they joked and drank, they fought; and while the women they met might retain them longer than intended in a town, a failed romance could hasten their departure. This represented a stage in the masculine life cycle, and partly explains why early modern journeymen showed such contempt for those amongst their number who married and established households. The reasons why authorities and guilds encouraged journeyman travel varied: it provided a structure for flexibility in local labour markets, as Ehmer shows, while German states, in particular, saw it as a force for technical competence and the diffusion of information about production methods.[30]

The *cursus honorum* represents the experience of many artisans, but the ideal was often no more than that. It assumed that a substantial proportion of journeymen could eventually set up as masters. A ratio of journeymen to masters of about two to one would mean, according to Garden, that most journeymen might realistically be seen as masters in waiting. Where the ratio was much less favourable, he found it difficult to see them as other than workers.[31] The ratio was deteriorating in many trades during the eighteenth century, though the problem was not new, and young men were less willing to submit to the master's authority in the household if they doubted that one day they would themselves be in his position.[32] Master-journeymen ratios alone can in any case conceal an effective dualism: Farr's research on Dijon between 1550 and 1650 uncovered a core of masters and chosen journeymen (most probably from better-established local families) surrounded by a much greater number of transient journeymen with few hopes of progress.[33] The core journeymen were far more likely to be the sons of masters, whether in the same trade or another. As Restif de Bretonne observed of the possibly extreme case of the eighteenth-century Parisian printing trades, 'workers never become masters ... masters beget masters, just as journeymen beget journeymen'.[34] Furthermore, the classic life-course categories of journeyman and master were increasingly unsatisfactory ways of capturing the nature of artisanal work structures even before industrialization. An urban economy of stable household workshops is a valid picture for many trades and, indeed, for many towns, but the networks of interlocking production units and sub-contracting that characterized production in an increasing range of trades made it hard to identify employers and masters. The dividing line between master and employee in the eighteenth-century English metal trades was imprecise: a journeyman in the Sheffield cutlery trades might be an outworker in one perspective but he possessed his own workshop, forge and tools.[35] Sonenscher's portrayal of the myriad of methods of employment, contract and payment in eighteenth-century French artisanal production is only the most graphic illustration of this problem.[36] One consequence was that the vision of the artisan making a whole product, on which interpretations of his moral as well as aesthetic significance partly rested, described a figure very much in decline.

The second defining feature of the artisanal ideal was *the fundamental role of the household in craft production*. The small workshop where the master shared his craft skills with his apprentices and journeymen, the intimacy of the residential setting, and the prospects of advancing along the artisanal life cycle were expected to create a stable unit in which the quality of production, labour discipline and harmonious workplace relations could all be assured. Masters were accorded patriarchal authority within their own expectations and, in the case of Germany, by legislation, which frequently gave them regulatory powers over apprentices, journeymen and servants. Yet it is important not to overstate the introspection of this artisanal workshop. Many trades were carried on away from the workshop – the building trades inevitably, but also others where the artisan worked on the customer's premises. This practice was widespread well beyond the aristocratic circles examined in Coquery's essay (Chapter Five). Farge has found that even where workshops remained the location of production in eighteenth-century Paris they were far from introspective places, with personnel and activities spilling out on to the street and into the neighbourhood, and a great deal of coming and going.[37] The classic model also requires a stability of journeyman employment which was often absent. It was perhaps only when the assumed life cycle was broken, and married journeymen became the norm, that there appeared in the nineteenth century a stability of employment which had been rarer before, thus misleading observers into believing that it was a natural component of artisanal production.[38] However, the evidence of journeyman turnover, as indeed of poor workplace relations, tends to come from capital cities and archives such as police and judicial records, and until we know more about employment conditions in smaller towns caution is needed before assuming that the integrated and relatively stable artisanal household was a mere fiction.

The artisanal life cycle and the artisanal household implied that marriage and mastership coincided. Marriage, admission as a master, and setting up on one's own account were intertwined in the guild regime, but the way they survived its abolition suggests that guild constraints were less crucial than were the practical imperatives set by the enterprise itself.[39] A master needed a partner who would contribute to the productive and selling side of the business, while also running the often complex household of apprentices, journeymen and servants. The reasons for an aspirant master artisan to marry a woman who herself came from a small enterprise background were thus related to practical as well as status concerns.[40] Marriage provided not only the couple needed by the enterprise but often the capital as well, for the bride's dowry frequently brought resources essential for setting up.[41] As Perrot has concluded for eighteenth-century Caen, the real obstacles to mastership were not juridical but financial, less the need to be accepted by the guild than to have the capital to acquire a workshop, its equipment and raw materials, and to pay for the master's licence.[42] Hence the high level of married masters both inside and outside the guild system, and hence the frequent coincidence of marriage and setting up.[43] The removal of guild restrictions did not produce radical change, especially in Germany, where rights of residence and hence of marriage were often bound up with master's status even after the abolition of guilds. Even the need for marital respectability, central to the guild sense of honour, repeated itself outside the guild order, for a respectable family was necessary for the longer-term credit without which few nineteenth-century small enterprises could survive. The ideology of the artisanal household as the unit of production which could best assure social stability and moral control was a continuing dimension of the artisanal ideal. The protests of male master tailors against Turgot's short-lived edicts abolishing French guilds in 1776 were couched in just these terms, but then so, too, were those of nostalgic German master artisans a century later.[44]

The position of apprentices and journeymen in artisanal households has received more attention than that of the masters' own children, perhaps reflecting assumptions about the relationship between family and workshop. Ehmer's study of household structure, work patterns and continuity of enterprise amongst Austrian artisans in the eighteenth and nineteenth centuries, however, led him to conclude that the fundamental artisanal institution under the *ancien régime* was not the family but the guild. The transfer of the business between generations was constrained by guild requirements, while the enterprise's labour needs were met by apprentices, journeymen and servants rather than by the master's wife or sons. These sons served their apprenticeship in other households, and travelled and were employed elsewhere as journeymen. If they themselves became masters, then it was generally by inheriting influence and resources rather than craft and workshop. The nineteenth-century trend in the more prosperous trades towards family labour and inheritance was, in this perspective, less the continuity of an older tradition than the adoption of bourgeois conceptions of family and property in the face of the declining influence of the guild and the declining presence of living-in workers.[45] Ehmer's findings may reflect a distinctively Central European pattern – a similar process can be found amongst Swedish artisans whose development was influenced by German customs and structures.[46] These distinctions were rarely as clear elsewhere in Europe, but continuity of enterprise was unusual, and it was common practice to send sons into other small enterprise households rather than retain them in the family business.⁴⁷ The presence of apprentices and journeymen in households from which sons were departing, as amongst Turin tailors in the first half of the eighteenth century, indicates that absence of work was not the reason. The aim was rather to secure more solid alliances for both the son and the family.⁴⁸

The craft as personal and social identity, the third feature of artisanship to be considered, rests on the belief that the artisan was above all else defined by his work. For many artisans throughout the period this proposition was indeed true, though by the late nineteenth century the fact that craft pride was less important than business skill made it an embattled identity. Nevertheless, even as relatively prosperous a master artisan as Joseph Brown, the Birmingham toolmaker interviewed by de Rousiers in the 1890s, 'still flatters himself that no smith in Birmingham can turn out better or quicker work'.⁴⁹ However, that pride was not fundamental to the identity of this successful petit-bourgeois businessman, as it had once been for a good proportion of master artisans and their aspirant journeymen. Even for those journeymen whose ambitions of mastership were waning, the defence of craft honour distinguished them from others who worked for wages but shared no such identity. Attachment to the craft included a very practical commitment and affection. Ménétra's opposition to guild restrictions did not preclude a passion for the craft itself: he kept a book in which to sketch workmanship that he admired, and visited shops in search of ingenious procedures and designs.⁵⁰ Many an artisan was bound up in the craft in which he had grown to adulthood, its fraternities and associations, its values and culture – including the culture of work.

However, the assumption that the craft was a pre-eminent phenomenon which shaped lifetime experience need not always have been the case. One critical assessment is offered by Farr in his essay (Chapter Three), using cultural analysis to question the priority accorded to work in the study of artisanal identity. The pre-eminence of the craft raises further problems. We have already seen the way trade labels were imposed that did not necessarily match the experience of those whom they embraced, nor necessarily reflect the multiplicity of occupations of many who were recorded as having only one. In eighteenth-century Bayeux, for example, the craft was not always the main source of income of those identified as artisans: land investment was often more important and could treble or quadruple what was earned from the workshop.⁵¹ Or consider the case of two independent woollen-weavers in Halifax at the end of the seventeenth century. The value of the craft goods which they left – primarily looms and cloth – were similar. Yet one was a wealthy yeoman with substantial property in livestock and farm goods, while the other was a simple weaver with only three cows and an old horse. The meanings of the trade for each would, in all likelihood, have been rather different: for one it was a mere adjunct to agricultural income; for the other it was the centre of his livelihood.[52] Beyond these examples lay the myriad of individual and family occupations which belie the simplicity of an individual occupational designation.

The assumed link between guild organization and artisanal identity has already been questioned. York boasted a multiplicity of metalworking guilds in the fifteenth and sixteenth centuries including smiths, locksmiths, blacksmiths and cutlers, yet Norwich at the same time required all these trades to be part of one smiths' guild. Guild constitutions were thus more a function of municipal policy than of craft identities. The 50 guilds at Lübeck, the 28 at Strasbourg and the 15 at Basel in the same period suggest that similar political influences were at work.[53] None of which is to deny the significance of guilds for their members; merely to stress that guilds and artisanal identities did not always coincide. The meanings of guilds to their members in seventeenth-century London and Norwich would have been real, but not necessarily occupational, because members' actual occupations were often very different from that of the guild or company to which they belonged.[54] Being an artisan in general was perhaps what mattered – being part of that workshop-based middling group of early modern and industrializing society. Such an identity would have varied with size of town, because a critical mass of members of a craft would have been needed for identity to grow through daily social relations. The bulk of Swedish guilds had no more than five or six members in the early nineteenth century, and this was probably true of many small-town guilds in Germany and Austria. When a guild was so small, what would it have meant for personal identity? This was particularly significant in Sweden where, as Edgren shows in his essay on Malmö (Chapter Seven), artisans' civic identity was primarily as members of the larger burgher estate in a setting where the civic role of guilds was minimal, and in social terms as part of a wider craft community rather than as a member of a specific trade. It was not only in small towns that the broader artisanal social world was what mattered. It made sense for artisans to construct a network of family ties across different crafts and economic interests which could provide a multiplicity of business connections, bases for credit, sources of influence and destinations for children.[55] Here is one explanation of a strategy present amongst European artisans since the middle ages. Farr's study of Dijon between 1550 and 1650 shows that while ties within the corporation were relevant to the town's artisans, they were less important than those within the wider artisanal community. The evidence of children's marriages, marriage witnesses and the choice of godfathers indicates that solidarity amongst masters transcended that within individual guilds.[56] Diversification within the middling milieu was an essential artisanal strategy in Europe during the period covered by this book.

The concept of the artisan was closely bound up with masculinity, and *craft and gender* is therefore the fourth feature of artisanship to be considered. The term 'artisan' came almost by definition to apply to men alone, and artisanal practice was built on relationships in which male authority prevailed. The artisanal career was predicated on the male life cycle: apprenticeship, several years of travelling, and the establishment of a stable independent enterprise which often drew on a wife's dowry. It took no account of the greater obstacles for working women in travelling alone, nor of interruptions to the career through child-bearing.[57] The definition of an artisanal trade increasingly required that it be male and those dominated by women were relegated to non-artisanal status. Consider the fate of seamstresses, milliners and dressmakers, rarely designated as craft mistresses or craft workers by contemporaries or, indeed, until recently by historians. When the knitting frame was introduced into Germany in the sixteenth century it made stocking-knitting easier and faster, but it was argued that it was complicated and that only men could do it, and women were prohibited from practising the craft on the grounds that they were unskilled.[58]

Yet women clearly played a role as employers and as skilled workers in artisanal trades, especially outside the guild system but even within it. The primarily administrative and political character of guild and occupational records meant that women's exclusion from formal public life in effect excluded them from a dominant section of the archives.[59] The female workforce fell largely outside the guild system, which provided the template on which the artisanal world was modelled. Yet, however much these sources understate women's role in craft production, both as mistresses and as family or hired labour, it would be wrong to see the masculine nature of artisanship as a function of representations alone. The role of women in craft enterprise remained restricted and came to be increasingly so, primarily permitted in terms of each individual's relationship to a male artisan.

The number of women masters was always small and the range of their trades narrowed during the early modern period. Fifteenth- and sixteenth-century Cologne was an exception, not only admitting women to most craft guilds but also boasting women's guilds for yarn-makers, gold-thread spinners, and silk-makers, whose members owned workshops and hired apprentices in their own right. Yet they were excluded from the city's political bodies.[60] Women masters were nowhere endowed with the social and political status of male craft masters. In her essay on eighteenth-century Nantes (Chapter Eight), Musgrave shows the existence of female sections of guilds which admitted women other than widows. Their position was firmly subordinate, and indeed these female sections were established largely to ensure control of their activities. Eighteenth-century Paris saw female trades such as linen-makers, seamstresses and hairdressers, with their own formal statutes and apprenticeships, and with organizations similar to male corporations, but they lacked their public role and often their legal status. The linen-drapers stand out as a well established mercantile guild that distanced itself from household production, whereas the more marginal seamstresses' guild was limited to very specific garments and was primarily a device to enable the state to channel clandestine female tailors into a guild for the purposes of their regulation.[61]

Female access to the formal practice of artisanal crafts was being restricted from at least the fifteenth century in much of Europe, with increasingly precise definitions of which women (wives, widows, daughters, servants) were acceptable in artisanal trades and the tasks they might perform. The pace and detail of change varied with place and trade, but the trend was unmistakable, aided by the fact that even where women controlled enterprises their lack of political voice made them powerless to protect their position.[62] Women's opportunities were greatest where guild control was weakest, including newer trades without institutional structures. Yet regulations would be introduced as a trade's economic importance grew, resulting in male control becoming established. Settled masters and orderly households were the goal, as in sixteenth-century Augsburg, for the Reformation ideal of the family was in many ways moulded from the patriarchal household of craft masters.[63] The right of masters' widows to continue the business was increasingly constrained by restricting the period during which they could operate the enterprise, their right to hire journeymen and train apprentices, and by denying them a place in guild deliberations. The outcome was that few widows were able to maintain the business for any length of time: the loss of the craft master, the paying of debts left at his death, the pressures of maintaining the family, and the obstacles which guilds set in her way meant that, as Musgrave confirms in her essay on Nantes, most abandoned any attempts to continue.[64]

Nevertheless, widows could continue the business because in the household-based workshop system they were active in the enterprise. Unlike the men's learning of a trade, that of women was essentially informal and intermittent. The master's wife would not only be responsible for the household, but often also for the purchase of raw materials, retailing, and supervising journeymen when her husband was absent, while in smaller enterprises or at busy times she (along with daughters and female servants) would play a part in the production process itself, especially in the subsidiary tasks involved in preparation or finishing.[65] Women tended to move in and out of the labour market or the household enterprise according to demand and their own availability, which reinforced their subordination in a sphere which rested on both systematic training and a masculine life cycle. This relationship to artisanal small enterprise continued in the nineteenth century, for the structures which sustained it went far beyond the juridical framework of guilds.[66]

The masculinity which increasingly defined artisanship was not a single cultural phenomenon. In Wiesner's study of early modern Germany, she contrasts the patriarchal masculinity of the master artisan with the boisterous masculinity of the travelling journeyman. Journeymen's anxieties over their rights often expressed themselves in an aggressive concern to minimize the involvement of any women other than the master's wife, and as they designated as dishonourable workshops or towns in which women were habitually employed, so the masculine nature of their own position became stressed, as Farr underlines in his essay.[67] This identification of women with dishonour, and later with undercutting, reinforced the sense that artisanship was a male phenomenon, and as new skills appeared, so they were appropriated by male workers.[68] The early nineteenth-century struggles of journeymen artisans against deteriorating conditions in their trade entrenched it still further, often targeting the use of female and domestic labour.[69] Without gender it is not possible to grasp the construction of what it meant to be an artisan, while without women's labour it is often equally impossible to explain the survival of artisanal production.

The urban world of the artisan

The ideals of artisanship were best articulated in small communities where craftsmen and their institutions were densely woven into the structures of a traditional urban world. The classic case was those German centres of up to 15 000 inhabitants which Mack Walker has called 'home towns'. The guildsmen's role in town government, their ability to protect themselves against wider state authorities, and the absence of severe outside economic pressures combined to create a stable and introspective world for the master artisans who dominated it.[70] Their control of access to economic, residential and burgher status meant that 'the German home towns in the peaceful years from Westphalia to Austerlitz created an unusual mechanism to repel intrusion, and that mechanism became the basis of its political and social life'.[71] Displaying what has been called their 'visceral xenophobia',[72] such towns represent the classic embodiment of the guild world, hierarchical, patriarchal and stable, defending its local autarkic economies against both immigrants and the intrusive state. Walker's conclusions are predicated on the special character of such towns, and care must be taken not to picture them as the artisanal norm before industrialization. Indeed, idealizations tend to rest on small-town settings, while our picture of a more unstable and conflictual world is drawn from large – especially capital – cities. Future research must bring urban size more clearly into focus, by reflecting on the tendency to inflate conflict that is inherent in the sources used for large-town studies, and by exploring the tensions within smaller urban centres, especially those involved in production for wider markets.

The historic centrality of artisans to the urban world was not limited to these classic small-town environments, for artisans were in many ways intrinsically rather than just incidentally urban. This is not to say that there were no rural artisans. On the contrary, they were a necessary part of the agrarian environment, providing basic services to both consumers and farmers, while industrial production in the countryside – often linked to the part-time labour of families otherwise engaged in agriculture – was a substantial part of manufacturing output in pre-modern Europe that grew in importance with proto-industrialization. Nevertheless, artisans were historically bound up with the definitions and meanings of towns well before industrialization generalized the urban experience. Manufacturing might be excluded from the countryside by law, as in Prussia or Sweden before the reforms of the early and mid-nineteenth century respectively. These restrictions sought to concentrate within towns the conflicts which surrounded industry and trade, to protect the peasantry from such tensions and, in the case of Sweden, to support the Crown's efforts to strengthen the urban presence. In fact, artisanal activity was never as absent from the countryside as such statutes might lead one to expect. Even where rural artisans were permitted, the exclusive rights of urban guilds would be protected, as with the laws which forbade country shoemakers round Bologna from selling in the town itself, while allowing their urban counterparts to sell higher-quality shoes in the countryside. As Bologna's shoemakers insisted throughout the eighteenth century, crafts were an urban activity and had to be defended as such.[73]

From the middle ages European towns had often come to be constituted around the juridical framework which gave the right to exercise a trade or profession to its guilds and corporations. These were a town's most basic institutions, bridging the private world of economic interests and the public sphere of urban government and ritual. Municipal governments saw in guilds a means to police both production standards and labour and market relations. Guild formation was encouraged in much of North-Western Europe from the fifteenth century onwards, and through the early modern period they became intermediary bodies for the regulation of prices, quality and the behaviour of apprentices and journeymen.[74] In much of Europe artisanal corporations sustained the civic as well as the social order of the town, often constituting the institutional core of urban government. They were not generally dominant – though where the urban patriciate was rooted in local mercantile activity they might maintain their corporate links – but they frequently played a significant role. In early sixteenth-century Augsburg, for example, a careful equilibrium of patrician and guild power saw the guilds overwhelmingly dominant on the Great Council, but in a more careful balance with wealthy patricians on the more influential Small Council.[75] Those may have constituted the peak years for guild masters in German urban politics for, as Friedrichs shows in his essay, artisanal representation on city councils declined during the later sixteenth and seventeenth centuries. In much of early modern Europe the right to practise a trade as a master, the rights of citizenship, and the rights of residence were closely related, especially in the German areas of Central Europe, and even within the rather different institutional setting of England until the seventeenth century a man often had to be free of the city to practise an independent economic activity. In Sweden, all masters had to be members of a guild and burghers of their town, though, as Edgren argues in his essay on Malmö, it was as burghers and craftsmen rather than as guildsmen that they took their place in the political and social order. Burghers, burgesses, freemen – urban citizenship and becoming a craft master were closely intertwined in many countries, though less so in France, where political rights were not required for mastership, and where urban elites were far more exclusive.[76]

The close, if diverse, relationships between masters, guilds and municipality meant that early modern artisans took their place within ritual presentations of the civic order. With their patron saints and churches, their place in public ceremonials and processions, their banners and symbols in the public display of the civic order, artisans and their guilds were fundamental to the proclamation of a town's identity. Nevertheless, this ritual activity only offered very mediated representations of the artisanal presence, representations which often varied with local and state configurations of political power. It is a point best appreciated by looking at those towns whose artisanal presence in civic representations was most muted yet where we know the artisanal urban presence to have been strong. Thus, as Edgren's essay shows, the symbolic representation of eighteenth-century Swedish towns was essentially as military entities rather than as collections of corporate crafts, while in seventeenth-century Turin the image which the city presented of itself in public ritual was that of the unitary municipality, and it was only when the municipality's authority was undermined in the eighteenth century that a fragmented set of crafts came to dominate public displays.[77]

Deventer in the Dutch Republic, with a population of 8000 in the 1780s and an economy that was based mostly on local markets, offers a parallel to the German home towns. Deventer's public life was built around its artisanal, trading and merchant guilds, whose civic and social prominence was symbolized by the town's ancient and beautiful *Gildehuis*. Yet a study of the political upheavals there during the 1780s reveals the fissures which could appear within such a seemingly traditionalist community when sections of the town's artisans and traders were drawn into more distant economic relationships.[78] It was, nonetheless, master artisans' role in pre-modern small-town life which led social catholics in the later nineteenth century to see their urban functions as crucial to their importance. 'In former times', claimed Oscar Pyfferoen with respect to Belgium, 'they have constituted the rampart of our communal liberties.'[79] Here was a historic commitment to liberty, independence and communal rights that had made small artisanal enterprise an essential force in small-town communities. Artisans were most successfully part of a very particular urban world, one undermined by the rapid urbanization of the nineteenth century.

The guild and its heritage

For Théophile Funck-Brentano at the end of the nineteenth century, guilds were part of an integrated and cohesive social order. 'Under the old social organisation the journeymen cared about the master, the master about the corporation, the corporations about one another, and all about the prosperity of the industry and trade of the city.'[80] For a proportion of urban artisans, for the town's identity and structure of order, and for those in the nineteenth century (amongst them Funck-Brentano) seeking to understand and sometimes to remedy the problems of artisans, the guild was a central institution. For many artisans in Central and Eastern Europe, including those in Hungary, examined in Bácskai's essay (Chapter Ten), the guild remained the idealized solution to the problems they faced in nineteenth-century economies.[81] However much we qualify the match between the guild ideal and the nature of artisanal relations, the presence of guilds was fundamental to much of the European urban economy between the sixteenth and eighteenth centuries. Guilds were variously concerned with economic regulation, the transmission of the craft, the control of labour, the representation of masters in legal and other disputes, the maintenance of quality, and the protection of their own members against market competition, but they might also embrace religious activity, urban ritual and social welfare. The functions guilds discharged varied with place and with time, but guilds played a significant role in urban economic and social relations and in the nature of the urban polity.

The literature on guilds in the period covered by this book becomes increasingly concerned with dissolution and decline, which is hardly unexpected given their withering in early modern Britain, their abolition in France in 1791 and in Belgium four years later, in Sweden in 1846 and 1864, and in the German territories during a more extended process which began in the Rhineland in the 1790s and ended only in 1859 in Austria and finally in 1869 in Mecklenburg. However, abolition is not necessarily the most helpful notion through which to approach guild history. Guilds were for a long time rather neglected by historians, as if their role and meaning were self-evident, and only in recent years has their character been opened to new questioning.[82] The decline and abolition of guilds suggest that they were a concrete and fairly uniform phenomenon, whereas they are better approached as heterogeneous institutions about which it would be wrong to assign priority to the economic over the religious, political and social dimensions. The emphasis on decline and abolition also implies that once guilds had been removed then corporate discourse would wither along with them, and it above all privileges the juridical structure as the crucial determinant of artisanal experience. The fact that guilds were a juridical entity does not mean that juridical abolition would eliminate what they constituted. The economic world of artisanal production should perhaps be seen as a structure of power, relationships and practices in continuing change and development which were linked to, but not entirely derived from, one single but unstable idea: the guild or corporation.

The limited resistance offered by master artisans and their guilds when the *loi d'Allarde* abolished corporations in France in March 1791 should be seen in this light. The lists of grievances (*cahiers de doléances*) produced by corporations in the winter of 1788-89 seemed to take their continued existence for granted, and were primarily concerned with the elimination of specific abuses.[83] This did not prevent masters denouncing the exclusive privileges of others, and by 1791 commitment to the Revolution, allied with a sense that little could be done to save corporations, led masters to acquiesce in the *loi d'Allarde* or, as Pontet finds in her essay on Bordeaux (Chapter Six), to offer only token resistance. Yet few thought that the outcome would be the end of all regulation in the world of work, as Hirsch has shown for the merchants and artisans of Lille who acquiesced in the end of their corporations.[84] War and the pace of revolutionary change on the one hand, and the commitment of public authorities to apply the laws on the other, speeded the transition from the old order to economic freedom. In any case, municipal authorities often saw corporations' exclusive privileges as incompatible with the rights of man established by the Revolution.[85] The ideological hold of guild ideas amongst artisans, as amongst the wider public, had been weakened over time by Turgot's temporary abolition in 1776,[86] the decrees abolishing privilege in August 1789, and the new discourse of equality and the rights of man. Masters acquiesced not only because municipal revolutions had given them access to new public functions, but because their own agendas had moved on, and the defence of guilds seemed part of an older set of concerns and an alternative discourse.

Guilds were of course varied institutions, not least in the balance in each body between trade jurisdiction, religious confraternity, and welfare and sociability. The autonomy enjoyed by guilds as urban institutions varied particularly across national systems. Whereas in France and, to a lesser extent, Germany guilds experienced a real existence independent of the state, notwithstanding the powers such as tax-collecting exercised on the latter's behalf, in other countries such as Sweden, Austria and Hungary they depended substantially on the state for their authority and activities, and Ehmer's essay shows how guilds in eighteenth-century Vienna were progressively strengthened as an instrument of state policy. Guilds could thus operate in different ways to serve distinct political interests. Friedrichs reveals in his essay the way guilds could provide craft masters with a continuing mode of intervention in city politics, whether to support the municipality, invoke outside authority, or protest on their own account. These relative levels of autonomy would warrant further study, but even the question of the presence or absence of guilds remains ambiguous in its meanings. The British case might seem clear, with guilds there of declining significance from the seventeenth century, but the survival of autonomous artisanal expectations and their expression in terms of trades rights and protection puts the British experience much closer to the French than institutional comparisons would lead one to expect. This conclusion is confirmed by Berlin's essay (Chapter Four), in which he shows not only the unevenness of the decline of London companies, but also the vigorous assertion of artisanal rights notwithstanding that decline. Or consider the case of Turin, where for most of the seventeenth century guilds barely existed, other than on paper, and displayed little concern for trade affairs, only to flourish in the early eighteenth century when changing configurations of municipal and ducal power left merchants and wealthy artisans in search of an institutional base with which to defend their interests now that the municipal council and commercial courts were no longer available to them. The rise of the guilds, above all in the 1720s and early 1730s, was the consequence. Even if caution is needed before extending Cerutti's conclusions beyond the specific political configuration of Turin, her analysis forces us to consider more closely the meanings we attach to the presence or absence of guild institutions.[87]

The transition from guilds was about far more than juridical abolition, not least because the ideals of artisanship were under various pressures during the eighteenth century and, indeed, before.[88] Five main processes can be identified. First, journeymen's access to mastership was becoming more difficult in many towns as the resources needed for independence increased and as larger workshops emerged in many trades. Second, there were increasing tensions between journeymen and masters, not least as the former sought to advance their interests as workers, albeit workers in a world they still perceived through a corporate lens.[89] This was not just a question of capital cities – as is revealed for France by journeymen's strikes in various trades in Lyon and Nantes, linked to increasing journeyman association around *compagnonnages* and religious confraternities.[90] A study of strikes in German cities between 1780 and 1805 shows considerable conflict between journeymen and masters over the denial of journeymen's traditional rights and their sense of honour and trade morality.[91] Wages may have been more explicitly prominent in Britain and honour in Germany, but journeyman action there, as in France, was articulated in terms of traditional rights and the defence of the community of the trade against misguided masters. Third, there was increasing inequality amongst masters which generated conflict between the guild elite and the mass of masters, and which also highlighted divergent economic interests, especially in those trades where sub-division and sub-contracting undermined the notion of an independent master, and where merchants and elite masters who controlled production had significant control within the guild itself. The division between London Company officers and the small-master 'yeomanry' explored by Berlin in his essay is an extreme version of such tensions. This was linked to the fourth process weakening the artisanal ideal: the increase in production outside the guild system. This was particularly marked in Paris, where *faux ouvriers* operated illegally or in privileged places such as the faubourg Saint-Antoine, or the *sauvetés* in Bordeaux described in Pontet's essay. These producers were not part of some underworld economy, but tied into the same systems of sub-contracting and wholesaling as the legitimate masters.[92] Suburban growth could carry masters beyond guild jurisdictions, as Berlin shows in his essay on London. These processes were thus weakening guilds in most countries, but particularly so in France, and Pontet's essay presents a picture of a corporate system that had become an increasingly empty shell, by 1791 if not by 1789. This leads us directly to the fifth process, the challenges to guilds from states which came increasingly to see them as restrictive institutions. Bácskai's essay on Hungary, for example, shows increasing government permission for production outside guilds, including permitting many journeymen to set up independently. In many countries the state was asserting its monopoly of jurisdiction and policing while also concerned to augment industrial production, yet it often restrained its hand in the eighteenth century, torn between a concern to increase production and anxieties about disorder, and guilds therefore survived best in countries such as Germany and Austria where their capacity to ensure order in the urban community and the world of work were most appreciated. The Prussian government, in fact, operated distinct policies in later eighteenth-century Minden-Ravensburg, undermining guilds in the linen-weaving export trade but sustaining them in industries such as shoemaking which served local markets.[93] The intensity of these five processes was increasing in the face of both economic expansion and intellectual challenges to the corporate organization of the economy and the society of orders within which guilds were located.

Yet guilds could serve as well as inhibit capitalist production. Merchants would often be members of guilds – as in French textile towns such as Lyon, Lille and Orléans – for they provided an excellent mechanism for the control of production quality, markets and credit, and for the disciplining of workers and sub-contracting small masters.[94] Merchants outside the system of production, and seeking to evade guild restrictions, might find their institutional base in chambers of commerce and their equivalent, but those who were *marchands-fabricants* involved in putting-out work often used guilds and corporations as an effective instrument of domination. Although France offers the best known examples, textile guilds in eighteenth-century Austria were being used in a similar fashion by larger merchants and putters-out.[95]

Guilds need not, therefore, be antithetical to capitalist development,[96] and in any case the free market was something which had to be constructed, rather than what was left when guilds were abolished. The history of artisanal production has too often been seen as a conflict between guild regulation and freedom of enterprise, whereas various regulations on the opening of enterprises and on relations of production often remained after the juridical abolition of guilds, as in both Germany and Sweden. The regulation of production should remain on the historian's agenda even after one particular form of it – the guild – has been removed. Concentration on guild abolition in analyses of artisanship also privileges the polarity of guild Europe and non-guild Europe at any one moment. Britain is too often left out of the picture or presented merely as the non-corporate comparison. Guilds did decline precociously in Britain, and their impact by the beginning of the eighteenth century was limited. Guilds and companies ceased to represent the interests of master artisans as larger-scale merchant capital prevailed, especially with respect to London's livery companies. Municipal government lost interest in guilds as an instrument of industrial and social policing, and legitimate small masters were by the early eighteenth century left with few friends. Guilds often survived, but were less and less relevant to artisanal production and to the crafts whose names they bore.[97] Yet the absence of effective guilds from eighteenth- and early nineteenth-century Britain did not weaken the power of artisanal ideals, which were now forcefully articulated by journeymen and small masters through the language of 'the trade'. The roots of London artisanal radical ideas of the early nineteenth century lay in eighteenth-century craft concern for status, honour, independence, the rights of the trade, skill as property, and control over access to the trade.[98] The idea of a craftsman's skill as his property was clearly presented during the struggles around 1814 against the repeal of the apprentice clauses of the Statute of Artificers, but its components were familiar eighteenth-century artisanal notions of the rights of its members to regulate a trade, to control entry, to exclude those without an apprenticeship, and the right of a journeyman to pass on the trade to his son.[99] When the Luddites not only proclaimed the right to live by one's trade but insisted on their right of inspection to judge the quality of workmanship, they were operating within the corporate idiom.[100] Berlin's essay emphasizes the power of the idea of the search in artisanal defence, whether for illegal or poor-quality goods. Exclusion was a central tenet of eighteenth-century artisanship. In Britain there were corporate ideals without corporations, ideals which were appropriated by the journeymen rather than their masters.

The transition from the corporate idiom within artisanal production cannot, therefore, be captured by precise moments of abolition, and the decline of guilds did not in itself remove the corporate influence in artisanal affairs. By the end of the nineteenth century the guild ideal had been appropriated by the political right but this had not always been the case. The trajectory of artisanal ideas in the changing situation of the nineteenth century was partly dependent on which sections of the artisanal world were the carriers of artisanal and corporate ideals during the century of European industrialization, and which sections detached themselves from it. In England and France, journeymen and very small masters often joined to resist the changes imposed by merchant capital and larger masters, defending the trade against those deemed to be undermining its traditional ideals. Elsewhere, especially in Germany and Austria, the line was more firmly drawn between those with master's status and those without it. In this context, the distinct legal character of mastership meant that even small masters distinguished themselves from journeymen.

In Germany, the protracted process of abolition both sustained the guild idea and signified its continuing strength. Whereas in France the *loi d'Allarde* had been rapidly followed by the *loi Le Chapelier* outlawing all associations within a trade, the establishment of freedom of occupation in Germany was usually accompanied by continuing masters' organizations. In Prussia and various other states, the links between practising an artisanal trade and the rights of urban residence allowed municipalities to control the number of masters. *Innungen* were quasi-public associations of master artisans which extended specific features of guilds into the era of 'trade freedom', such as representing trade interests, supervising the education of apprentices, and undertaking a variety of benevolent activities. Guilds did not so much disappear in Germany, they were simply stripped of many of their legal functions, and the continuing existence of associations straddling the public and private spheres left the corporate option in place for successive generations of German artisans seeking an alternative to the decline that they feared. In Sweden all masters were required to join the *Hantverksföreningar* (craft associations) after guilds had been abolished, and these became central to masters' mobilization, even after 1864 when they ceased to be compulsory.[101] The continuation of *Innungen* was one institutional reason why masters became the main bearers of the corporate ideal in Germany. When King Frederick William IV passed briefly through Cologne in 1841, some 600 of the city's master artisans presented him with a picture of their misery and demanded the return of closed, compulsory corporations with formal masters' examinations.[102] In 1848, the restoration of guilds, limitations on factory production, and the banning of non-artisans from selling artisanal products all shaped master artisans' demands at the Frankfurt Artisans' Congress. Yet, alongside guild restoration, they called for progressive income and property taxes, free state education, and financial support for journeymen on the *Wanderjahre*.[103] Although corporate demands were important, so too was the commitment of the same master artisans to democratic political and social change, as can be seen at the local level, for example in Düsseldorf, where demands for guild restoration came not from those trades closest to the traditional guild ideal, but from the tailors, shoemakers and cabinetmakers who were in structural crisis, and who were also prominent in the democratic reform movement. It is unwise simply to assume that guild ideas implied reactionary politics, at least in the first half of the nineteenth century.[104]

In France and Britain, however, it was journeymen artisans and the most vulnerable of small masters who protested in terms of traditional artisanal ideals while masters, without distinct legal or institutional status, operated increasingly as small businessmen untempted by the restrictions of the artisanal past. French *compagnonnages* survived guild abolition, and their eighteenth-century past shaped their traditions, modes of organization and concern to maintain the quality of artisanal labour, while their continued role in the *tour* gave them the basis of a control over the labour market which they used to put pressure on masters considered to be working against the interests of the trade. Although there were limits to their influence, the *compagnonnages* constituted significant bearers of corporatist ideas that were to weaken only during the July Monarchy. The early French labour movement bore witness to these continuing preoccupations: associationism, the defence of artisanal rights, the notion of the trade as property, the commitment to self-organization, and the power of a moral critique of productive relations.[105] Sewell has argued that journeymen carried the corporate idiom into the French labour movement of the first half of the nineteenth century, insisting that the ideals of artisanship constitute the basis of work and productive relations.[106] Although doubt has been cast on the continuities of journeymen's organizations, and although not all artisanal trades saw workers take the classic eighteenth-century craft as their point of reference,[107] the continuing power of the corporate tradition remains necessary to an understanding of the French working-class movement through to 1848. The same argument has already been seen with respect to Britain, where the enforcement of customary artisanal practice was left increasingly to journeymen, and where their defence of the custom of the trade is reminiscent of artisanal discourse elsewhere in Europe. The trade had to organize itself to defend the artisan's property in his skill, to limit competition, and to maintain the standards of the craft. In the movements of eighteenth-century protest, and in the radical and trade associations of the first half of the nineteenth century, we see the capacity of the artisanal ideal to sustain protest and shape its discourse, even where its institutional base had long been marginalized.[108] The fate of the term 'artisan' is in this context illuminating. Whereas the German *Handwerker* and the French *artisan*, unless further qualified, tended in the nineteenth century to refer to independent workshop owners, the word 'artisan' in English came to be used as the standard designation for an employed skilled worker, whether in craft or heavy industry.[109]

The specialization and marginalization of artisanal production

One reason for the appeal of artisanal ideals as the nineteenth century progressed was that they offered an alternative vision of the relations of production, at a time when artisans were subjected to pressures that differed in intensity, and ultimately in character, from those previously experienced. It is to these pressures that this final section will turn. The impact of industrialization on artisanal production was neither as rapid nor as catastrophic as once thought. As growth rates flatten, as the significance of hand labour and traditional skills is stressed, as large-scale production is shown to have been limited to very specific sectors for much of the nineteenth century, so we have come to understand the centrality of small units of production to industrialization.[110] There were distinguishable national patterns. In Britain, Belgium and France, where the industrialization process during the first half of the nineteenth century rested on consumer goods and on a producers' goods sector in which scale did not significantly increase outside the iron industry, the proliferation of small units of production (including artisanal workshops) was a central feature of industrial change. In Germany, where industrialization occurred later, centred more on capital goods, and was more closely linked to large-scale finance, workshop production declined more strikingly, even when integrated within commercial structures as elsewhere in Europe. The proportion of artisanal and industrial employees in enterprises with five employees or fewer fell from 60 per cent in 1882 to 31 per cent in 1907.[111] The French pattern provides a striking contrast. Caron concluded that before 1914 'small and even very small industrial enterprises still played a very important part in the French industrial structure'.[112] Revisionist analyses have concluded that small and medium-sized enterprises, far from being a brake on industrial development, were the medium through which France's industrialization took place, assuring the major part of economic growth through to the 1930s.[113] Yet the contrast amongst Europe's first industrializers should be kept in perspective. Even in Bochum, the diversification which followed initial heavy industrial development created opportunities for small producers whose numbers grew substantially.[114] In Britain, the fact that the great adoption of steam engines came in the last quarter of the nineteenth century signals something which historians have progressively come to appreciate: in most industries production grew not through factories and capital-intensive technology, but through the proliferation of various forms of dispersed production.

Nevertheless, small units of production expanded in the context of changing relations of production which subordinated artisanal and other small producers within networks controlled by large capital. Subordination took different forms, from the debasement of artisanal trades into impoverished domestic outworkers on the one hand, to the uses of orders and credit to make artisanal workshops dependent on merchant capital on the other. The growth of domestic outwork in a small number of artisanal trades has long been recognized as a central feature of industrialization, though one primarily concentrated in textiles and certain consumer goods industries such as shoemaking (not exclusively so, however, as the impoverished nail- and chainmakers of the English Black Country remind us). The more pervasive impact of industrialization on artisanal production, however, lay in the enmeshing of notionally independent workshops within structures dominated by large capital, through intensive sub-division of tasks and sub-contracting on the one hand, and dependence on large capital for credit and orders on the other. These two related processes ensured that the independence of workshop production weakened.

These changes were not new, for much of the tension in eighteenth-century workshop relations and many of the inequalities amongst masters followed trends towards sub-contracting, specialization and dependence. The pervasiveness of guild discourse before industrialization can conceal the organizational heterogeneity within workshop production. The apparent fragmentation into independent workshops in eighteenth-century Paris, Sonenscher has argued, conceals a complex division of labour by which individual processes were sub-contracted to different masters yet integrated within a single process of production.[115] Garden's exploration of the silk-weaving *fabrique* in Lyon on the eve of the Revolution revealed a more intensive structure of subordination. He found 350 *marchands-fabricants* manufacturing on their own account or putting work out to others, and 6000 master artisans (*façonniers* in fact, receiving orders from merchants) described as 'working for others'. 'The *façonniers*', Garden concluded, 'in spite of their title as masters, in spite of their workshops and their ownership of the looms which were their tools of production, were entirely dominated by the *marchands-fabricants*.'[116] The forces that subordinated workshop masters to larger capital were growing in eighteenth-century industry, especially in the larger cities.[117]

It was the nineteenth century, however, that saw the effective integration of workshop producers into a capitalist structure characterized by merchants and factors sub-contracting to a myriad of notionally independent masters. The tailoring trades of London and Paris in the 1830s and 1840s provide the most striking case, as they faced the competition of ready-made clothing produced for the new wholesale merchants. Many masters retreated into taking work on sub-contract, while others became no more than domestic pieceworkers. Independence was now an illusion for all but a minority of bespoke tailors and viable sub-contractors.[118] Tailoring was an extreme case, but one with parallels in most trades producing goods for wide-scale consumption, and a similar process can be seen in the metal trades where more traditional craft skills survived better. Indeed, Berg has stressed the vitality and adaptiveness of eighteenth-century small-scale production in Birmingham and Sheffield, before the polarization between large and small producers and the intensification of dependence in the following century.[119] During the nineteenth century, there and in other metalworking towns such as Solingen and Paris, the production of hardware, cutlery, locks and such goods expanded rapidly through intense subdivision of trades under the organizing influence of merchants and factors.[120] These were the link between artisanal producers and national and international markets, and as such industries grew so workshop owners became increasingly dependent on individual merchants for orders, credit for raw materials, and even advances to pay wages. The burden of fluctuation and competition was thrown on to the workshop master, something especially attractive in trades where vulnerability was created by either rapid changes in fashion, for example ribbon-weaving in Saint-Etienne, or unstable demand, as in Birmingham's gun production. In Birmingham's gun quarters fairly prosperous workshops proliferated under merchant organization.[121] The master gunmaker gathered in his warehouse the parts made in independent workshops – items such as locks, barrels, sights and triggers – before distributing them to other specialized workshops for assembling. The artificial flower trade of Paris may have been a humbler affair, but it was similarly organized: the 'rosemaker', for example, did nothing but assemble the flower in his warehouse from the products of 10 separate independent workshops.[122] By the late nineteenth century, few artisanal sectors had escaped the intervention of a third party between producer and consumer, or escaped the links of financial dependence, which is why the Belgian economist Victor Brants described most small producers as no more than 'vassals of their suppliers of raw materials'.[123]

Independence for most artisans became an ambition rather than a reality, even for those who succeeded in opening their own business, and the growing constraint affected not only established industrial centres but also primarily administrative and commercial towns, as studies of Edinburgh and Toulouse have made clear.[124] The changes in a growing number of consumer goods industries during the first half of the nineteenth century subsequently became generalized to most sectors. The constraints of capitalist competition transformed the conditions of work, the nature of specialization, and the economic prospects and freedom of most artisanal trades in industrial Europe. For as long as dispersed production was capable of meeting market demand in terms of quantity, price and quality, then there were few incentives for large entrepreneurs to enter directly into production themselves, and the factory sector need advance only slowly in artisanal trades. Some explanations for the survival and, indeed, expansion of small-firm production have stressed its economic adaptability and responsiveness, and its ability to change techniques and products rapidly to meet the demands of differentiated markets. According to Sabel and Zeitlin, those districts of industrial Europe characterized by 'flexible specialization' generated social institutions to provide protection from the extremes of the market, greater financial stability, and a stimulus to continuous innovation.[125] The specific character of certain nineteenth- and twentieth-century regions might be described in these terms, but care must be taken not to overgeneralize from their example. The opportunities for most artisanal producers in the nineteenth century brought with them constraints which seemed increasingly at odds with the ideals of artisanship through which many still assessed their situation. An unexpected shift in idealization seems to have occurred in the historical literature – from artisanal production as rigid and restrictive in the guild era to artisanal production as flexible in the industrial era. Each refers only to very specific situations.

The transformations outlined above were not evenly spread through the urban hierarchy. In small towns whose economies rested on local and agrarian customers, small artisanal producers survived under the protection of those local markets. Agricultural depression and the rural exodus were weakening their position by the 1870s, but the real challenge came from changes in transport and marketing which exposed them to regional, national and often international competition. Many a shoemaker or tailor saw his customers seek more fashionable goods elsewhere, while many a wheelwright or smith was left with repair work as expensive items ceased to be purchased locally. In contrast stood the industrial towns, resting on the varieties of dispersed production described above, where many of the greatest pressures on artisanal producers were experienced, as sub-contracting and specialization in sub-divided trades progressively reshaped production in sections of industries such as textiles, clothing, furniture and metalworking. Yet nowhere did artisanal production seem to come under such intense pressure as in the capital cities of industrializing Europe, where specialization and sub-contracting threatened and ultimately destroyed the independent workshop in most trades producing for consumer markets. Bourillon's essay (Chapter Eleven) on the renewal of the artisanal economy in one Parisian quarter during Haussmann's reconstruction depicts its two striking characteristics: first, its resilience in the face of a process of demolition and reconstruction which is classically thought to have destroyed small enterprise; second, the basis of that resilience in specialization, sub-division of tasks, and sub-contracting which served only to make the renewed artisanal sector far more dependent on large capital. The proliferation of small-scale production in Europe's great nineteenth-century capital cities was a sign not of their backwardness,[126] but of their ability to adapt to the requirements of industrial capitalist economies within distinctive metropolitan contexts. Although a stress on responsiveness and flexibility might indicate economic adaptability and progress, it does not preclude the fragmentation, sub-division, competition and decline of skill which characterized so much artisanal production in London, Paris or Berlin. Indeed, the very problems faced by artisans were a result of the dynamism of their metropolitan economies.[127]

The nineteenth century saw a marked advance in artisanal specialization, and for most artisans that sub-division of tasks amongst workshops was the basis of their ultimate decline. Specialization was not new, and Coquery's essay shows a minutely detailed specialization of trades serving aristocratic consumers in late *ancien régime* Paris. These were the pinnacle of the luxury trades which continued to serve elite markets during the following century, and the importance of taste and quality production has been particularly stressed as a factor in French industrialization, following the lead of the political economists Adolphe Blanqui and Michel Chevalier, who had returned from the 1851 Great Exhibition convinced that the quality of France's craft industries was the only basis for future success in world markets.[128] In reality, the specialization which principally characterized artisanal trades was not that serving elite consumption, but the minute sub-division of processes through putting-out and sub-contracting to serve mass consumption. This was the basis, for example, of the decline of the major part of the clothing, shoemaking and furnituremaking trades of London and Paris into debased artisanal producers, who often ended up in the casualized sectors designated 'sweated' by the close of the nineteenth century. The challenges to artisans were felt earliest and most acutely in capital cities, where a combination of large-scale demand and over-supply of labour provided entrepreneurs with opportunities for specialized sub-division of labour. Cottereau concluded that later nineteenth-century Parisian artisans had little to do with 'the former corporate and artisanal crafts'. As he explained, 'the bases of the division of labour ... were no longer modelled on the distinctions between crafts' for large-scale capitalist organization constantly redistributed work so as to place as much as possible with those who lacked a recognized skill: labourers, women, the young, the old. As a result, 'workers no longer dreamed of reconquering a lost artisanal paradise'.[129]

The debased and sweated trades described by Mayhew in mid-century and Booth at its end for London, or by Du Maroussem in his meticulous studies of Parisian artisanal production in the 1890s, were not the totality of artisanal production even in those cities.[130] In many metal trades, small-scale engineering, and indeed much consumer goods production, sub-division and sub-contracting did not destroy a workshop structure with master and workers so much as lock it into the competitiveness and dependence considered above. With whatever variations by sector or by kind of town, by the end of the nineteenth century specialization and sub-division characterized the economic experience of most artisans, and this contributed to an inescapable narrowing of their social and cultural role. Indeed, as Kooij demonstrates in his essay (Chapter Twelve) on the industrial development of Groningen in the late nineteenth century, the town's dual economy and dual labour market ensured that the modern sector acquired a higher status, leaving crafts as a low-paid and low-status sector especially dependent on immigrant labour. If this dualism may have been linked to the specific features of Dutch economic and urban development in that period, the picture offers a stark version of the declining position of artisans - masters and journeymen - in western economies. Artisanal production throughout Europe moved over the long term from a central position in the urban economy to one of marginality. That the artisanal sector could still renew itself, especially in the assembly and maintenance of new industrial products and techniques, is confirmed by the substantial growth of electricians, bicycle makers and repairers, car mechanics, plumbers and so on from the 1890s. Older artisans might sometimes move into these new areas to defend their business position. Paul Marcelin shifted the emphasis of his family's tinsmith's business in Nîmes to new plumbing skills at the end of the nineteenth century to take advantage of the growing demand for domestic bathrooms.[131] Yet new trades such as these remained on the margins of production and of artisanship. The artisanal sector declined less in scope or numbers during the nineteenth century - that was a matter for the century that followed - than in significance and profitability. If the pressures could all be found in previous centuries, their scale and pervasiveness within industrial capitalist economies was new. As Bácskai's essay shows, the decline of artisans as a group, in wealth and status as well as in economic significance, was not limited to industrializing Europe. The increase in imported mass-produced goods, rural craft production, and competition within the crafts, combined to undermine the position of Hungarian artisans in the nineteenth century.

The increasing specialization and marginality of artisanal production distanced it ever further from the ideals bound up in the notion of the artisan. As that distance grew, resistance to economic pressures and to the changing relations of production often drew strength from the idealized vision of artisanship, and used it as a yardstick against which the present was to be judged and found wanting. The myth was the construction of neither a few politically minded artisans and intellectuals in the first half of the nineteenth century, nor of social conservatives anxious about industrial society towards its end. The idealization of the artisan was the product of a continuing tension between the character of artisanal production, and the terms in which artisans themselves came to make sense of their experience, often by developing a conception of the past through which to engage critically with the present. The historical vision of artisanship thus engaged in a continuing relationship with the changing urban economy, the changing urban political and juridical order, and the changing relationship between artisans and the urban social and symbolic order. We have seen how civic and religious processions often sought symbolically to represent the place of artisans within the urban order of early modern Europe. By the latter half of the nineteenth century, however, the ritual presence of artisanship had declined into iconographic vestiges, the symbols and images of crafts carved into town hall façades or displayed in local museums. They were drawn into the historical construction of place which became so important in late nineteenth-century Europe. Artisanship was thus appropriated not only for conservative critiques of modernity as we saw at the beginning of this essay, but also for the assertive construction of proud new civic and local identities. Such identities may have been a far cry from the artisanal world which they claimed to evoke, but they remained a powerful witness to the place of artisanship in the urban imagination.

Notes

1. This was recognized by the Belgian social catholic economist Victor Brants, who felt that 'historical memories' gave to the identification of craft industry its traditional physiognomy. V. Brants, *La petite industrie contemporaine*, Paris: Lecoffre, 2nd edn (1902), p. 2.
2. A. Babeau, *La ville sous l'ancien régime*, Paris: Didier, 2nd edn (1884), p. 45. For a German example of nostalgic construction of artisanship, see E. Mummenhoff, *Der Handwerker in der deutschen Vergangenheit*, Leipzig (1901). I am grateful to Christopher Friedrichs for this reference.
3. A. Babeau, *Les Artisans et les Domestiques d'autrefois*, Paris (1886), p. v. Examples of those using Babeau include Brants, *La petite industrie*, p. 27ff. and L. Rivière, 'La notion des classes moyennes', in *Les Classes*

Moyennes dans le Commerce et l'Industrie. XXIXe Congrès de la Société internationale d'économie sociale, Paris (1910), p. 4.
4. F. Husson, Artisans français. Les menuisiers. Étude Historique, Paris: Marchal & Billard (1902) and Artisans français. Les serruriers. Étude Historique, Paris: Marchal & Billard (1902).
5. J. Ehmer, 'The artisan family in nineteenth-century Austria: embourgeoisement of the petite bourgeoisie', in G. Crossick and H.-G. Haupt (eds), Shopkeepers and Master Artisans in Nineteenth-Century Europe, London: Methuen (1984), pp. 212-14.
6. La Réforme sociale, 48, 1904, pp. 61-2; T. Ericsson, 'Cults, myths and the Swedish petite bourgeoisie, 1870-1914', European History Quarterly, 23, 1993, pp. 245-6.
7. L. Edgren, 'Craftsmen and political consciousness in Sweden 1850-1900', unpublished paper to First Conference on the Nordic Middle Classes, Århus, 1993.
8. C.E. Schorske, Fin-de-Siècle Vienna. Politics and Culture, New York: Alfred A. Knopf (1980), pp. 66-72. In Britain, the linking of aesthetic and social visions, though with different political implications, can be found in the Arts and Crafts movement and in William Morris and his followers.
9. For the perspective on the petite bourgeoisie held within some social catholic and Le Playist circles, see Geoffrey Crossick, 'Metaphors of the middle: the discovery of the petite bourgeoisie 1880-1914', Transactions of the Royal Historical Society, 6th series, vol. 4, 1994, pp. 251-79.
10. Le Comte de Boissieu, 'Le moteur éléctrique et l'industrie à domicile dans la région lyonnaise', La Réforme sociale, 55, 1908, p. 656.
11. A. Champetier de Ribes, 'Le petit commerce, la petite industrie, et la question de l'apprentissage', La Réforme sociale, 60, 1910, pp. 451-2.
12. See for example O. Mélon, L'Ordre social et les classes moyennes, Ghent: Librairie de l'Association Belge pour le développement des Classes Moyennes (1933).
13. Quoted by B. Zarca, L'Artisanat français du métier traditionnel au groupe social, Paris: Économica (1986), p. 54. For Vichy perspectives on artisans, see ibid., pp. 54-63, and S.M. Zdatny, The Politics of Survival. Artisans in Twentieth-Century France, New York and Oxford: Oxford University Press (1990), pp. 128-53.
14. L. Roper, The Holy Household. Women and Morals in Reformation Augsburg, Oxford: Clarendon Press (1989), p. 55.
15. S.L. Kaplan, 'Réflexions sur la police du monde du travail, 1700-1815', Revue historique, 261, 1979, p. 71.
16. J.-L. Ménétra, Journal of My Life, with an introduction and commentary by D. Roche, New York: Columbia University Press (1986), p. 286.
17. Quoted by Edgren, 'Craftsmen and political consciousness'.
18. For an example, see the description of eighteenth-century cabinetmaking as the background to its early nineteenth-century decline, with the earlier period portrayed as one of integrated production and high levels of broadly based skill, and a craft community uniting those within the trade: L.S. Weissbach, 'Artisanal responses to artistic decline: the cabinetmakers of Paris in the era of industrialization', Journal of Social History, 16, 1983, p. 68.
19. Ménétra, Journal of My Life, p. 18.

20. Zarca, *L'Artisanat français*, p. 9ff.
21. Ménétra, *Journal of My Life* and A. Perdiguier, *Mémoires d'un compagnon* (original ed. 1854-55), Paris (1964).
22. J. Rancière, 'The myth of the artisan: critical reflections on a category of social history', in S.L. Kaplan and C.J. Koepp (eds), *Work in France. Representations, Meaning, Organization, and Practice*, Ithaca: Cornell University Press (1986), pp. 317-34. Quotation p. 321. For the fuller statement of Rancière's challenge, see his *The Nights of Labour: the Workers' Dream in Nineteenth-Century France*, Philadelphia: Temple University Press (1989).
23. H. Swanson, 'The illusion of economic structure: craft guilds in late medieval English towns', *Past and Present*, **121**, 1988, pp. 29-48; S. Cerutti, *La ville et les métiers. Naissance d'un langage corporatif (Turin, 17e-18e siècle)*, Paris: Éditions de l'École des Hautes Études en Sciences Sociales (1990).
24. C.-J. Gadd, *Självhushåll eller arbetsdelning? Svensk lant- och stadshantverk ca 1400-1860*, Gothenburg: Göteborgs Universitet (1991), p. 424.
25. D. Keene, 'Continuity and development in urban trades: problems of concepts and the evidence', in P.J. Corfield and D. Keene (eds), *Work in Towns 850-1850*, Leicester: Leicester University Press (1990), p. 10.
26. On attempts to enumerate small enterprises in France, Germany and Britain, see G. Crossick and H.-G. Haupt, *The Petite Bourgeoisie in Europe 1780-1914: Enterprise, Family and Independence*, London: Routledge (1995), pp. 39-40. On the ideological nature of occupational classification, A. Desrosières and L. Thévenot, 'Les mots et les chiffres: les nomenclatures socioprofessionnelles', *Économie et Statistique*, 110, 1979, pp. 49-65; E. Higgs, 'The struggle for the occupational census, 1841-1911', in R. Macleod (ed.), *Government and Expertise: Specialists, Administrators and Professionals, 1860-1919*, Cambridge: Cambridge University Press (1988), pp. 73-86.
27. J.R. Farr, *Hands of Honor. Artisans and Their World in Dijon, 1550-1650*, Ithaca and London: Cornell University Press (1988), esp. ch. 3, pp. 122-49; Cerutti, *La ville et les métiers*, esp. p. 67ff.
28. M. Gribaudi and A. Blum, 'Des catégories aux liens individuels: l'analyse statistique de l'espace sociale', *Annales ESC*, 1990, pp. 1365-402.
29. F. Lenger, *Sozialgeschichte der deutschen Handwerker seit 1800*, Frankfurt: Suhrkamp (1988), p. 18ff.
30. For a good comparative survey, see U.-C. Pallach, 'Fonctions de la mobilité artisanale et ouvrière - compagnons, ouvriers et manufacturiers en France et aux Allemagnes (17e-19e siècles)', *Francia*, 11, 1983, pp. 365-406.
31. M. Garden, 'Ouvriers et artisans au XVIIIe siècle. L'exemple lyonnais et les problèmes de classification', *Revue d'histoire économique et sociale*, 48, 1970, pp. 41-2.
32. Roper, *The Holy Household*, pp. 33-6.
33. Farr, *Hands of Honor*, p. 138ff. For a similar picture in early nineteenth-century Malmö, see L. Edgren, *Lärling Gesäll Mästare. Hantverk och hantverkare i Malmö 1750-1847*, Lund: Universitetsförlaget Dialogos (1987), p. 400.
34. Quoted in J. Materné, 'Chapel members in the workplace: tension and

teamwork in the printing trades in the seventeenth and eighteenth centuries', in C. Lis, J. Lucassen and H. Soly (eds), *Before the Unions. Wage Earners and Collective Action in Europe, 1300-1850, International Review of Social History Supplement 2* (1994), p. 54.

35. M. Berg, *The Age of Manufactures 1700-1820*, London: Fontana (1985), pp. 279-80.
36. M. Sonenscher, *Work and Wages. Natural law, politics and the eighteenth-century French trades*, Cambridge: Cambridge University Press (1989).
37. A. Farge, *Fragile Lives. Violence, Power and Solidarity in Eighteenth-Century Paris*, Cambridge, MA: Harvard University Press (1993), pp. 104-30.
38. See the regularity and stabilities of the Marcelin tinsmith's enterprise in Nîmes during the second half of the nineteenth century, as recalled by Paul Marcelin in his 'Souvenirs d'un passé artisanal', *Les cahiers rationalistes*, 253, 1968, pp. 33-72, though many of their journeymen still became small masters.
39. For the role of the family in nineteenth-century small enterprise, see Crossick and Haupt, *The Petite Bourgeoisie in Europe*, pp. 87-111.
40. On July Monarchy Paris, for example, see A. Daumard, *Les bourgeois de Paris au XIXe siècle*, Paris: Flammarion (1970), p. 132. And see the comments recorded by Paul de Rousiers in the 1890s on why the son of a Birmingham toolmaker was marrying the daughter of a clockmaker in the town, in his *The Labour Question in Britain*, London: Macmillan (1896), p. 35.
41. For examples see E. Musgrave, 'Women in the male world of work: the building industries of eighteenth-century Brittany', *French History*, 7, 1993, p. 37; B. Angleraud, 'Les boulangers lyonnais aux XIXe-XXe siècles (1836 à 1914). Une étude sur la petite bourgeoisie boutiquière', doctoral thesis, University of Lyon 2, 1993, vol. 1, pp. 136-45.
42. J.-C. Perrot, *Genèse d'une ville moderne. Caen au XVIIIe siècle*, Paris: Mouton (1975), p. 340.
43. See for example Edgren, *Lärling Gesäll Mästare*, p. 400 for Malmö.
44. J.G. Coffin, 'Gender and the guild order: the garment trades in eighteenth-century Paris', *Journal of Economic History*, 54, 1994, p. 782; S. Volkov, *The Rise of Popular Antimodernism in Germany. The Urban Master Artisans, 1873-1896*, Princeton: Princeton University Press (1978), p. 119.
45. J. Ehmer, 'The artisan family in nineteenth-century Austria: embourgeoisement of the petite bourgeoisie?', in G. Crossick and H.-G. Haupt (eds), *Shopkeepers and Master Artisans in Nineteenth-Century Europe*, London: Methuen (1984), pp. 195-218.
46. E. Lunander, *Borgaren blir företagare. Studier kring ekonomiska, sociala och politiska förhållanden i förändringens Örebro under 1800-talet*, Uppsala: Acta Universitatis Upsaliensis (1988), p. 151.
47. M. Mitterauer and R. Sieder, *The European family: Patriarchy to partnership from the Middle Ages to the Present*, Oxford: Blackwell (1982), pp. 80-81.
48. Cerutti, *La ville et les métiers*, p. 170.
49. De Rousiers, *The Labour Question*, p. 5.
50. Ménétra, *Journal of My Life*, p. 73.

51. O.H. Hufton, *Bayeux in the Late Eighteenth Century. A Social Study*, Oxford: Clarendon Press (1967), p. 74.
52. J. Smail, 'Manufacturer or Artisan? The relationship between economic and cultural change in the early stages of the eighteenth-century industrialization', *Journal of Social History*, 25, 1991-92, p. 795.
53. Swanson, 'The illusion of economic structure', pp. 41-2; T.A. Brady, 'Economic and social institutions', in B. Scribner (ed.), *Germany. A New Social and Economic History: vol. 1, 1450-1630*, p. 267.
54. M. Pelling, 'Apprenticeship, health and social cohesion in early modern London', *History Workshop Journal*, 37, 1994, pp. 38-9.
55. For Turin, Cerutti, *La ville et les métiers*, p. 18; for eighteenth-century Birmingham, M. Berg, 'Women's property and the industrial revolution', *Journal of Interdisciplinary History*, 24, 1993-94, p. 250.
56. Farr, *Hands of Honor*, pp. 122-49.
57. A point made by M. Wiesner, *Women and Gender in Early Modern Europe*, Cambridge: Cambridge University Press (1993), p. 103.
58. M.E. Wiesner, 'Gender and the worlds of work', in Scribner (ed.), *Germany*, p. 215.
59. Swanson, 'The illusion of economic structure', pp. 39-40.
60. M. Wensky, 'Women's guilds in Cologne in the later Middle Ages', *Journal of European Economic History*, 11, 1982, pp. 631-50.
61. D. Garrioch, *Neighbourhood and Community in Paris 1740-1790*, Cambridge: Cambridge University Press (1986), p. 113; Coffin, 'Gender and the guild order', pp. 768-93.
62. Wiesner, *Women and Gender*, pp. 103-4. For a case study see I.K. Ben-Amos, 'Women apprentices in the trades and crafts of early modern Bristol', *Continuity and Change*, 6, 1991, pp. 229-37.
63. Roper, *The Holy Household*, pp. 47-9.
64. See for example ibid., pp. 49-53; Musgrave, 'Women in the male world of work', pp. 41-2.
65. O. Hufton, *The Prospect Before Her. A History of Women in Western Europe. Vol. 1. 1500-1800*, London: HarperCollins (1995), pp. 162-3.
66. Crossick and Haupt, *The Petite Bourgeoisie in Europe*, pp. 87-99.
67. M. Wiesner, '"Wandervogels" and women: journeymen's concept of masculinity in early modern Germany', *Journal of Social History*, 24, 1991, pp. 767-82.
68. Berg, *Age of Manufactures*, pp. 150-51; K. McClelland, 'Some thoughts on masculinity and the "representative artisan" in Britain, 1850-1880', *Gender and History*, 1, 1989, pp. 164-77.
69. B. Taylor, ' "The men are as bad as their Masters ...": Socialism, feminism and sexual antagonism in the London tailoring trade in the early 1830s', *Feminist Studies*, 5, 1979, pp. 7-40; J. Scott, 'Men and women in the Parisian garment trades: discussions of family and work in the 1830s and 1840s', in P. Thane, G. Crossick and R. Floud (eds), *The Power of the Past: Essays for Eric Hobsbawm*, Cambridge: Cambridge University Press (1984), pp. 67-93.
70. M. Walker, *German Home Towns: Community, State and General Estate 1648-1871*, Ithaca: Cornell University Press (1971).
71. Ibid., p. 136.
72. E. François, 'Immigration et société urbaine en Allemagne à l'époque moderne (XVIIe-XVIIIe siècle)', in M. Garden and Y. Lequin (eds),

Habiter la Ville. XVe-XXe, Lyon: Presses Universitaires de Lyon (1984), p. 39.
73. C. Poni, 'Norms and disputes: the Shoemakers' Guild in eighteenth-century Bologna', *Past and Present*, 123, 1989, p. 84ff.
74. See for example Swanson, 'The illusion of economic structure', p. 31.
75. Roper, *The Holy Household*, pp. 12-13.
76. C.R. Friedrichs, *The Early Modern City 1450-1750*, London: Longman (1995), p. 143.
77. Cerutti, *La ville et les métiers*, pp. 14-15.
78. W. Ph. Te Brake, *Regents and Rebels. The Revolutionary World of an Eighteenth-Century Dutch City*, Oxford: Basil Blackwell (1989).
79. O. Pyfferoen, 'La petite bourgeoisie d'après une enquête officielle à Gand (1)', *La réforme sociale*, 37, 1899, p. 292. See also L. Rivière, 'La notion des classes moyennes', in *Les Classes Moyennes dans le Commerce et l'Industrie. XXIXe Congrès de la Société internationale d'économie sociale*, Paris: Rousseau (1910), pp. 4-5.
80. Th. Funck-Brentano, 'Préface', in P. Du Maroussem, *La question ouvrière. vol. 2: Ébénistes du Faubourg St-Antoine*, Paris (1892), pp. 10-11.
81. See also Volkov, *Rise of Popular Antimodernism*.
82. See in particular the works of Cerutti, Kaplan, Sonenscher, Truant, Bossenga, Hirsch, Poni and Mackenney cited elsewhere in these notes.
83. B.F. Hyslop, 'French gild opinion in 1789', *American Historical Review*, 44, 1939, pp. 252-71.
84. J.-P. Hirsch, *Les deux rêves du commerce. Entreprise et institution dans la région lilloise, 1780-1860*, Paris: Editions de l'EHESS (1991), pp. 237-62.
85. M.P. Fitzsimmons, 'The National Assembly and the abolition of guilds in France', *Historical Journal*, 39, 1996, pp. 133-54.
86. Though Pontet's essay shows that the February 1776 edict was not applied in Bordeaux.
87. Cerutti, *La ville et les métiers*.
88. For a fuller discussion of this transition, see Crossick and Haupt, *The Petite Bourgeoisie in Europe*, pp. 16-37.
89. C.M. Truant, 'Independent and insolent: journeymen and their "rites" in the Old Regime workplace', in Kaplan and Koepp (eds), *Work in France*, pp. 137-8; S. Kaplan, 'Réflexions sur la police du monde du travail 1700-1815', *Revue historique*, 261, 1979, pp. 17-77. A recent survey of journeymen forms of association since the middle ages identifies an extensive history of collective organization and action by journeymen, which has implications for assessments of the extent of the eighteenth-century decline in relations: C. Lis and H. Soly, ' "An irresistible phalanx": journeymen associations in western Europe, 1300-1800', in Lis, Lucassen and Soly, *Before the Unions*, pp. 11-52.
90. Truant, 'Independent and insolent', pp. 131-73; D. Garrioch and M. Sonenscher, '*Compagnonnages*, confraternities and associations of journeymen in eighteenth-century Paris', *European History Quarterly*, 16, 1986, pp. 25-45.
91. J. Kocka, 'Craft traditions and the labour movement in nineteenth-century Germany', in Thane, Crossick and Floud, *The Power of the Past*, pp. 110-12.

92. S.L. Kaplan, 'Les corporations, les "faux ouvriers" et le faubourg Saint-Antoine au XVIIIe siècle', *Annales E.S.C.*, 43, 1988, pp. 353-78.
93. W. Reininghaus, 'Zünfte und Wirtschaftspolitik in den Territorien im Westen Deutschlands am Ende des Alten Reiches', unpublished paper to International Conference on the Eighteenth-Century Guild System in Theory and Social Practice, Halle, 1995.
94. G. Bossenga, *The Politics of Privilege: Old Regime and Revolution in Lille*, Cambridge: Cambridge University Press (1991), esp. pp. 131-67; idem, 'Protecting merchants: guilds and commercial capitalism in eighteenth-century France', *French Historical Studies*, 15, 1988, pp. 702-3.
95. J. Ehmer, 'Guilds in eighteenth-century Austria', unpublished paper to International Conference on the Eighteenth-Century Guild System in Theory and Social Practice, Halle, 1995.
96. For doubts about the necessarily restrictive character of guilds in an earlier period, see R. Mackenney, *Tradesmen and Traders. The World of the Guilds in Venice and Europe, c. 1250-c.1650*, London: Croom Helm (1987), pp. 78-130.
97. For a fuller discussion see Crossick and Haupt, *The Petite Bourgeoisie*, esp. pp. 23-6. Also G. Unwin, *Industrial Organization in the Sixteenth and Seventeenth Centuries*, London: Clarendon Press (1904); J.R. Kellett, 'The breakdown of gild and corporation control over the handicraft and retail trades in London', *Economic History Review*, new series, 10 (1957-58), pp. 381-94; L.D. Schwarz, *London in the age of industrialisation: entrepreneurs, labour force and living conditions, 1700-1850*, Cambridge: Cambridge University Press (1992), pp. 210-16.
98. I. Prothero, *Artisans and Politics in Early Nineteenth-Century London: John Gast and his Times*, London: Folkestone: Dawson (1979). See also the essay by Berlin (Chapter Four).
99. J. Rule, 'The property of skill in the period of manufacture', in P. Joyce (ed.), *The Historical Meanings of Work*, Cambridge: Cambridge University Press (1987), pp. 99-118.
100. Berg, *Age of Manufactures*, p. 279.
101. Edgren, 'Craftsmen and political consciousness'.
102. P. Ayçoberry, 'Histoire sociale de la ville de Cologne (1815-1875)', doctoral thesis, University of Paris 1, 1977, p. 317.
103. P.H. Noyes, *Organisation and Revolution: Working-Class Associations in the German Revolutions of 1848-49*, Princeton: Princeton University Press (1966), pp. 163-91.
104. F. Lenger, *Zwischen Kleinbürgertum und Proletariat. Studien zur Sozialgeschichte des Düsseldorfer Handwerker 1816-1878*, Göttingen: Vandenhoeck und Ruprecht (1986), pp. 150-87.
105. See for example C. Johnson, *Utopian Communism in France: Cabet and the Icarians 1839-1851*, Ithaca: Cornell University Press (1974); B.H. Moss, *The Origins of the French Labour Movement: The Socialism of Skilled Workers, 1830-1914*, Berkeley: California University Press (1976).
106. W.H. Sewell, Jr, *Work and Revolution in France: The Language of Labor from the Old Regime to 1848*, Cambridge: Cambridge University Press (1980).
107. L. Hunt and G. Sheridan, 'Corporatism, associationism and the language

of labor in France, 1750-1850', *Journal of Modern History*, 58, 1986, pp. 813-44; M. Sonenscher, 'Journeymen's migrations and workshop organization in eighteenth-century France', in Kaplan and Koepp, *Work in France*, pp. 74-96.
108. For the eighteenth century, J. Rule, *The Experience of Labour in Eighteenth-Century Industry*, London: Croom Helm (1981); for the early nineteenth century, Prothero, *Artisans and Politics* and C. Behagg, *Politics and Production in the Early Nineteenth Century*, London: Routledge (1990). For valuable comparative discussions of the artisanal contribution to labour movements, see F. Lenger, 'Beyond exceptionalism: notes on the artisanal phase of the labour movement in France, England, Germany and the United States', *International Review of Social History*, 36, 1991, pp. 1-23, and J. Breuilly, 'Artisan economy, ideology and politics: the artisan contribution to the mid-nineteenth-century European labour movement', in idem, *Labour and Liberalism in Nineteenth-Century Europe*, Manchester: Manchester University Press (1992), pp. 76-114.
109. G. Crossick, 'From gentlemen to the residuum: languages of social description in Victorian Britain', in P.J. Corfield (ed.), *Language, History and Class*, Oxford: Blackwell (1991), pp. 167-9.
110. See for example F. Caron, *An Economic History of Modern France*, London: Methuen (1979); N.F.R. Crafts, *British Economic Growth during the Industrial Revolution*, Oxford: Oxford University Press (1985); R. Samuel, 'The workshop of the world: steam power and hand technology in mid-Victorian Britain', *History Workshop Journal*, 3, 1977, pp. 6-72; Crossick and Haupt, *The Petite Bourgeoisie*, ch. 3, pp. 38-63; C. Sabel and J. Zeitlin, 'Historical alternatives to mass production: politics, markets and technology in nineteenth-century industrialization', *Past and Present*, 108, 1985, pp. 133-76.
111. G. Hohorst, J. Kocka and G.A. Ritter, *Sozialgeschichtliches Arbeitsbuch*, vol. 2, Munich: Beck (1978), p. 75; Lenger, *Sozialgeschichte*, p. 115.
112. Caron, *Economic History*, p. 164.
113. J. Bouvier, 'Une démarche révisioniste', in P. Fridenson and A. Straus (eds), *Le capitalisme français XIXe-XXe siècle. Blocages et dynamismes d'une croissance*, Paris: Fayard (1987), pp. 11-30. Also R. Roehl, 'French industrialization: a reconsideration', *Explorations in Economic History*, 13, 1976, pp. 233-81.
114. D. Crew, *Town in the Ruhr: A Social History of Bochum 1860-1914*, New York: Columbia University Press (1979), p. 17.
115. M. Sonenscher, 'Work and wages in Paris in the eighteenth century', in M. Berg, P. Hudson and M. Sonenscher (eds), *Manufacture in Town and Country before the Factory*, Cambridge: Cambridge University Press (1983), pp. 155-7.
116. Garden, 'Ouvriers et artisans', pp. 29-31.
117. For Britain, see M.D. George, *London Life in the Eighteenth Century*, London: Harmondsworth: Penguin (1966 edn), pp. 158-212; Berg, *Age of Manufactures*.
118. C.H. Johnson, 'Patterns of proletarianization: Parisian tailors and Lodève woolens workers', in J. Merriman (ed.), *Consciousness and Class Experience in Nineteenth-Century Europe*, New York: Holmes and Meier (1979), pp. 65-84; Taylor, '"The men are as bad as their

Masters ..."'; Scott, 'Men and women in the Parisian garment trades'.
119. M. Berg, 'Small producer capitalism in eighteenth-century England', *Business History*, 35, 1993, pp. 17-39.
120. See for example Behagg, *Politics and Production*; G.I.H. Lloyd, *The Cutlery Trades: An Historical Essay in the Economics of Small-scale Production*, London: Longman (1913); G.C. Allen, *The Industrial Development of Birmingham and the Black Country 1860-1927*, London: Cass (1966 edn); J. Gaillard, *Paris, la Ville (1852-1870)*, Paris: Honoré Champion (1977).
121. Allen, *Industrial Development*, pp. 116-19; M.J. Wise, 'On the evolution of the jewellery and gun quarters in Birmingham', Institute of British Geographers, *Transactions*, 15, 1951, pp. 57-72. See also J.R. Bailey, 'The struggle for survival in the Coventry ribbon and watch trades 1865-1914', *Midland History*, vii, 1982, pp. 142-8.
122. Gaillard, *Paris, la Ville*, pp. 438-9.
123. Brants, *La petite industrie*, p. 124.
124. E. Knox, 'Between capital and labour: the petite bourgeoisie in Victorian Edinburgh', Ph.D. thesis, University of Edinburgh, 1986, ch. 5; R.A. Aminzade, *Class, Politics, and Early Industrial Capitalism: A Study of Mid-Nineteenth Century Toulouse, France*, Albany: SUNY Press (1981).
125. Sabel and Zeitlin, 'Historical alternatives'. For an interpretation of London's industrial economy in these terms, see P. Johnson, 'Economic development and industrial dynamism in Victorian London', *London Journal*, 21, 1996, pp. 27-37.
126. Cf. G. Stedman Jones, *Outcast London. A Study in the Relationship between Classes in Victorian Society*, Oxford: Oxford University Press (1971).
127. See for example D.R. Green, *From Artisans to Paupers: Economic Change and Poverty in London, 1790-1870*, Aldershot: Scolar Press (1995); B.M. Ratcliffe, 'Manufacturing in the metropolis: the dynamism and dynamics of Parisian industry in the mid-nineteenth century', *Journal of European Economic History*, 23, 1994, pp. 263-328; Gaillard, *Paris, La Ville*.
128. W. Walton, 'Political economists and specialized industrialization during the Second French Republic, 1848-52', *French History*, 3, 1989, pp. 293-311.
129. A. Cottereau, 'Étude préalable', in D. Poulot, *Le sublime, ou le travailleur comme il est en 1870 et ce qu'il peut être*, Paris: François Maspero (1980), p. 70.
130. H. Mayhew, *London Labour and the London Poor*, New York: Dover (1968 edn); E.P. Thompson and E. Yeo (eds), *The Unknown Mayhew. Selections from the Morning Chronicle 1849-1850*, London: Merlin (1971); C. Booth, *Life and Labour of the People in London*, second series, 'Industry', London: Macmillan (1902); P. Du Maroussem, *La question ouvrière*, 4 vols, Paris: Rousseau (1891-94).
131. P. Marcelin, 'Souvenirs d'un passé artisanal', *Les cahiers rationalistes*, 253, 1968, pp. 42-4.

CHAPTER TWO

Artisans and urban politics in seventeenth-century Germany

Christopher R. Friedrichs

The city of Reval may seem like a curious place to begin our discussion of the role played by artisans in the political system of the seventeenth-century German city. Reval, after all, lay far from the German heartland: today it is better known as Tallinn, the capital of Estonia. Nor was its population entirely German, for the city had a sizeable underclass of Estonians and a sprinkling of other ethnic groups. Yet the politically active inhabitants of Reval were predominantly German and the city's institutions were entirely Germanic. Reval's political history, moreover, is exceptionally well recorded. Drawing on a rich and rather overlooked body of literature on the history of early modern Reval, we can preface our general observations about the role of artisans in German urban politics of the seventeenth century by seeing how the artisans of this particular city attempted to articulate and achieve their collective objectives.

But first we must know something of Reval's economic and political structure in early modern times. Reval was a thriving Baltic port in the seventeenth century, with a population that rose from about 5000 in the 1620s to almost 10 000 in 1708.[1] The city's economic importance derived largely from its role as a transfer point for the export of Estonian grain destined for Western Europe and for the import of salt, textiles and other products headed for the Baltic hinterland.[2] Like any such entrepôt, Reval had a small but wealthy merchant elite. In addition, however, the city had a large number of artisans who chiefly produced goods for local and regional markets or provided the services on which the urban economy depended.

The city's political organization reflected its economic structure.[3] As in most German cities, the artisans of Reval were organized into guilds, one for each craft. These *Ämter*, in turn, were grouped into two larger organizations, the *kleinen Gilden*. Of these, the St Kanutigilde was considered more prestigious: all its members were of German nationality and thus permitted to hold the rank of citizen. By contrast, the St Olaigilde included some *Ämter* whose members were non-Germans – mostly Estonians – and were thus generally barred from citizenship.[4]

There was a third *Gilde* in Reval – the *Grosse Gilde* of merchants. It was far more powerful than any of the craft organizations, for only members of the *Grosse Gilde* could serve as members of the city council, or *Rat*. As in most German cities, the council enjoyed enormous powers over the city's inhabitants, but at the same time it was answerable to the city's overlord – who, in the seventeenth century, happened to be the king of Sweden. For constitutional precedents and guidelines, Reval looked to German cities like Lübeck, but for the solution to specific problems Reval and its inhabitants had to turn to Stockholm.

Each craft in Reval had its own particular concerns. But looking at the seventeenth century as a whole, one can identify a number of objectives which most artisans – or, to be more precise, most artisan masters – collectively shared.[5] Many of their efforts were concerned with minimizing competition from outsiders. The artisans who were citizens of Reval constantly tried to block the activities of rival craftsmen who lived and worked in the countryside or even operated within the city itself in the protected precincts controlled by the Swedish governor. They were also determined to protect their markets by preventing the importation of manufactured goods similar to those they made themselves. The turners, for example, struggled against the importation of foreign spinning-wheels; the tinsmiths agitated against merchants who sold metalware from abroad.[6] The artisans also tried to limit the number of masters in specific crafts by imposing heavy financial burdens on applicants for membership, typically by requiring that new guild members host huge banquets – which many candidates simply could not afford. The tailors were even more direct; they tried to ensure that only sons of masters or men who married the daughters or widows of masters could accede to masterships themselves.[7]

However, the master craftsmen were not solely concerned with protecting the economic position of the particular trades to which they belonged. They also recognized a common set of interests which they all shared by virtue of their status as artisan masters. As a group they were eager to enhance their collective standing as members of the community. Though the artisan masters entertained no hopes of ever gaining membership in the city council, they wanted the two *kleinen Gilden* to enjoy parity with the *Grosse Gilde* in various aspects of communal administration, for example in the right to audit the council's financial records. They also wanted to make it easier for those artisans who gave up a craft and started to engage in trade to enter the ranks of the *Grosse Gilde*.

How did the artisans pursue these goals? Obviously the most common approach was the universal method of the day: to petition the city council of the day respectfully for an amendment to the guild statutes or for a new piece of municipal legislation that would embody the artisans' own goals. Of course this method was most likely to succeed when the council was sympathetic to the artisans – as when the craft masters were engaged in a struggle against rival artisans who were not citizens of Reval. Often, however, the master artisans' interests clashed with those of the merchant-dominated council, for example when they tried to prevent the merchants from importing finished goods into Reval. In situations like these the artisans were far less likely to get satisfaction from the council.

Yet the artisans had other means of pursuing their goals. The most obvious was to appeal over the councillors' heads directly to the ruler himself. In 1626 King Gustav Adolf came to Reval in person. The master artisans – represented by the leaders of the two *kleinen Gilden* – took advantage of the king's presence to appeal for protection from rival craftsmen. They also asked that the exclusive right enjoyed by some richer inhabitants to brew and sell beer be extended to them as well. The artisans gained some satisfaction on the first score, but none on the second. The king was recorded as having replied bluntly: 'The crafts will never flourish if a potter is also a brewer; a craftsman should be a craftsman and a brewer should be a brewer.'[8] Yet a royal visit was in any case a rarity. Normally the artisans appealed to the king's governor in Estonia, or they sent a delegation to Stockholm. The city council routinely protested against such autonomous actions by the artisans, but usually in vain: even if the artisans did not achieve their aims, they could almost always count on a respectful hearing from the governor or the royal councillors. For the Crown had an interest in not letting the council become all-powerful in Reval.

In 1659 the *kleinen Gilden* submitted an appeal directly to the king asking for protection against the city council's efforts to make admission to masterships easier. The royal authorities responded with vague promises to protect the artisans' traditional privileges. But the magistrates sternly reprimanded the artisans for sending delegates to Sweden and made it clear that they would try to force the guilds to lower their fees and accept new masters. A mounting cycle of demands and counter-demands finally led to a bloody confrontation in August 1662, when a group of artisans demonstrating at the city hall attacked the 20 soldiers whom the council had sent to restore order. The council immediately ordered the banishment of Hans Kemmerer, the head of the shoemakers' guild and master of the Kanutigilde who had emerged as the artisans' leader. But Kemmerer turned his disgrace to the artisans' advantage, for he sailed straight away to Stockholm to make the artisans' case before a royal commission. By and large this commission endorsed the council's policies – but it also absolved Kemmerer and allowed him to return to his position in Reval. Clearly the royal officials wanted to deny the council a one-sided victory over the city's artisans.[9]

The artisans certainly realized that their opportunities were heightened when the city's elite itself was divided. In 1670 a serious conflict erupted within the municipal oligarchy when the *Grosse Gilde* accused the city council of financial mismanagement and abuses of power. The artisans, speaking as usual through the two *kleinen Gilden*, quickly joined the fray. No doubt they reasoned that the city council, which was suddenly at odds with its customary allies in the merchant guild, might be more dependent than usual on support from the artisans. So the *kleinen Gilden* quickly issued a long set of demands, for more freedom from interference by the merchants, a more prominent role in civic ceremonies and the like. When the council delayed in responding, the artisans threatened to appeal directly to the king. They eventually did so, and the Crown granted many of their demands.[10]

Of course the artisans were not always united in their objectives. In the first place, there were persistent tensions between those artisans who belonged to the political community of Reval and those who did not, such as the craftsmen who worked in the protected enclaves of the Swedish officials. But conflicts also emerged among artisans who belonged to the two *kleinen Gilden*. Many artisans favoured the amalgamation of the two *Gilden* into a single, more powerful one – and this was indeed approved by the Swedish government in 1675. But ethnic tensions soon emerged. All members of the Kanutigilde were German, but many members of the Olaigilde were not. It soon became apparent that the German artisans intended to exclude all Estonians from the new amalgamated organization. Presumably sensing that the unified *Gilde* would function less cohesively if its ethnic composition were mixed, the city council tried to insist that all members of the Olaigilde, Germans and Estonians alike, must be admitted. This, however, was strongly resisted by the German artisans – and they prevailed. The Olaigilde gradually lost more and more of its German members and eventually disappeared altogether. But the Kanutigilde, strengthened by the addition of new members, continued to advocate effectively the interests of Reval's artisans – or at least those artisans who were of German extraction.[11]

Looking at the seventeenth century as a whole, one can see that the artisans of Reval achieved only a small portion of their long-term objectives – partly due to their own internal differences, and partly due to the opposition of other powerful groups. But it is also clear that throughout the century Reval's artisan masters were major players in the politics of their community. They fought with skill and tenacity for their objectives, and their aims and intentions always had to be taken seriously by the city's rulers. Artisans were central to the urban politics of seventeenth-century Reval.

Reval, of course, was only one of hundreds of cities within the Germanic regions of Central and Eastern Europe. In virtually every one of these cities, master artisans made up a majority of the adult males who enjoyed the status of *Bürger*, or citizen. It is obviously impossible to discuss the political history of these communities without paying some attention to the role of this significant group - a group whose members, moreover, were deeply conscious of their identity and collective interests. Yet the part played by artisans in the urban politics of seventeenth-century Germany is often overlooked. This is not because the artisans themselves are ignored by historians: the crafts and craftsmen - *Handwerke* and *Handwerker* - of early modern Germany have long been objects of intense interest.[12] The problem arises, instead, from the fact that the urban politics of early modern Germany is itself a somewhat neglected topic, and when it is examined, it is often treated in a way that obscures the role of artisans in the political system.

What exactly is meant by urban politics? It may be useful to begin by looking at one well known description:

> Politics arises out of conflicts, and it consists of the activities - for example, reasonable discussion, impassioned oratory, balloting, and street fighting - by which conflict is carried on. In the foreground of a study of city politics, then, belong the issues in dispute, the cleavages which give rise to them and nourish them, the forces tending toward consensus, and the laws, institutions, habits and traditions which regulate conflict ...[13]

These words come from a classic study of American urban politics in the mid-twentieth century. To a large extent, however, they would apply equally well to cities in early modern Europe - and certainly to German cities in the seventeenth century. The exact manifestations of urban politics in the seventeenth century were certainly different from those to which we are accustomed today. 'Balloting' generally played a smaller role then than it does in the modern city. But impassioned oratory and street fighting were important instruments of urban politics then as now. No doubt there was also much 'reasonable discussion' in seventeenth-century German cities, though this form of political conduct always leaves the fewest traces in any historical record. More importantly, however, the concept of urban politics as a pattern of structured conflict by which issues in dispute come to be resolved is as pertinent to understanding the seventeenth-century German city as it is to understanding any modern community.

Yet few historians of the early modern German city emphasize the character of urban politics as an ongoing system of conflict and its resolution. The reason for this lies, perhaps, in the way in which the exercise of power in the early modern city is understood. Studies of power in the modern city typically fall into 'elitist' and 'pluralist' approaches.[14] The 'elitist' approach assumes that communities are dominated by a ruling group and thus the central concern in any study of urban politics should be the composition, aims and methods of the ruling elite. The 'pluralist' approach, by contrast, assumes that decision-making is the result of the interplay of influences brought to bear by numerous groups within the community. This approach is closely linked to the perception of politics as the process by which different interest groups attempt to achieve their objectives through the governmental system. The pluralist approach is certainly central to the study of politics in modern urban communities. Yet it remains of little importance in treatments of the early modern German city. To an overwhelming extent, the 'elitist' approach predominates.[15]

There are important reasons for this. On first inspection, political power does appear to have been remarkably concentrated in German cities of the early modern era. The typical *Rat* or council embodied the concentration of very substantial executive, legislative and judicial authority in the hands of a small body of men who in many cases were selected by a process of co-optation with little or no popular input. A significant tendency in the writing of German history sees the Reformation of the sixteenth century as the last epoch of 'communal' influence on urban politics; thereafter, it is argued or implied, city governments became ever more oligarchical.[16] To be sure, it is not forgotten that there were periodic outbursts of popular discontent in the form of sudden violent conflicts between the citizens and their rulers. These conflicts between *Rat* and *Bürgerschaft* have been the object of intense study, and rightly so.[17] But generally these disputes are seen as temporary exceptions to the normal character of urban political life, in which the city councils ruled in an ever more authoritarian way, echoing absolutist governments in their determination to impose order and discipline on an unruly populace. The ordinary citizens, who, to a large extent, were artisans, are held to have declined in status from being participants in the urban political order to being mere subjects of the oligarchical magistrates.

Certainly any approach to urban politics which focuses on the composition of municipal elites would lend some credence to this argument. During the late middle ages, after all, craft guilds were frequently represented on city councils. Often the guilds held a formally allocated number of seats on the council; in other cities local custom ensured the selection of prominent guild masters to fill vacant positions.

In the course of the sixteenth and seventeenth centuries, however, artisanal representation on city councils steadily declined. Sometimes this was due to a formal constitutional change: in most of the imperial cities of Southern Germany, for example, guild representation on city councils was formally abolished by the Emperor Charles V between 1548 and 1552.[18] Elsewhere the extrusion of artisan members was a more gradual process; indeed, the direction of change was not always linear. In Hanover, for example, the composition of the council was actually broadened in the 1530s, during the upheavals of the Reformation era, so as to provide more seats for guild representatives. But in the seventeenth century the size of the council declined again and artisanal members were gradually squeezed out.[19] Certainly there was an unmistakable trend towards more oligarchical regimes: in many cities even merchants no longer dominated the councils as more and more power shifted to jurists and quasi-aristocratic rentiers.[20] Of course there were cities in which the system of guild representation persisted into the seventeenth century. Cologne was a notable example - but even in Cologne much of the council's political authority was in reality restricted to a small inner group of patrician mayors.[21] For in general it was only in the smallest urban communities - the 'home towns' whose political culture has been so perceptively described by Mack Walker - that the guilds continued to play a central role in the formal structures of local government.[22]

To focus on the composition of the political elite, however, can be misleading. It certainly diverts attention from the continued importance of artisans in the political life of German cities in the seventeenth century. The magistrates of many German cities in that era were wont to describe their powers in remarkably all-encompassing terms. But baroque formulations of absolute power had little to do with actual political practice in the German cities.[23] Ordinary citizens often raised objections to these overblown claims, and historians would be wise to do likewise. For in many ways the powers of the urban magistrates were subject to significant limitations.[24]

In the first place, there were limitations from above. All city councils ultimately stood under the authority of some higher ruler. The relative autonomy which had been enjoyed by many city governments in earlier centuries was steadily eroded by pressure from territorial princes, who wanted to turn urban governments into instruments of their own administration - a development which many magistrates themselves came to regard as unavoidable.[25] The absolutist rulers of the seventeenth century were often passionately interested in what was happening in specific cities. In October and November of 1683, for example, King Charles XI of Sweden personally chaired seven successive meetings devoted to the alleged misconduct of the mayor of Reval.[26] The electors, landgraves, margraves and other princes of Germany proper were no less interventionist in dealing with the cities of their realms. Nor were the so-called 'free imperial cities' - the *Reichsstädte* - spared this kind of pressure from above. Intervention by their overlord, the Holy Roman Emperor, was an infrequent occurrence, but when it happened it could result in significant constitutional changes.[27]

Yet the power of urban magistrates in seventeenth-century Germany was also significantly limited by pressures from below. No city council could rule without extensive cooperation from those over whom it governed. The first and most obvious reason was that the council normally lacked any means of coercion. In many German cities the citizens - who consisted chiefly of adult male artisans - had access to arms as members of a civic militia. This in itself gave the citizens some measure of power in their relations with the magistrates. Many city governments did employ a troop of professional soldiers, but the numbers were usually very small. When, in 1662, the city council of Reval sent 20 soldiers to deal with the rioting artisans, this may well have represented the entire number of professional soldiers at its disposal.[28] For practical reasons alone, the magistrates had to be careful not to antagonize the inhabitants of their community - especially the core group of householding artisans.

Limitations on the powers of the magistrates should not only be seen in negative terms, however, as obstacles to a presumed programme of increased social control on the part of the city's rulers. They should be seen, instead, as integral to the system of urban politics, a system in which different groups within the community attempted to pursue their objectives within a framework of traditional customs and expectations. In fact the pluralist notion of urban politics as the interplay of influences exerted by different interest groups within the community is fully applicable to the German city of the seventeenth century.

Naturally some groups could bring more influence to bear than others. The wealth and prestige of the overlapping elite groups - the patricians, merchants, jurists and the like - gave them substantial resources. People at the other end of the social scale - the floating population of house-servants, day-labourers, paupers and drifters, with no clear rights of residence - had virtually no social capital other than their moral claim on the charitable disposition of established residents. The artisans, however, did have significant resources to bring to bear in the interplay of urban politics.

In the first place, there was the sheer weight of numbers. In city after city, artisans and the members of their households made up by far the largest component of the inhabitants. But artisans also had the advantages derived from their legal status. Most adult male artisans who had achieved the level of master were also *Bürger*, or citizens. To be sure, the two groups were not completely identical. Some masters, as the case of Reval reminds us, were not citizens. And some male artisans who were citizens were nevertheless not masters – as in the occasional instances of journeymen who were granted civic rights without ascending to mastership. But generally the congruence between *Meisterschaft* and *Bürgerrecht* was high, and this meant that most heads of artisan households not only enjoyed the guaranteed right to live and work in the city but also held at least latent rights of participation in the political order.

Above all, however, the artisans' capacity to function effectively as interest groups within the city resulted from their extremely high level of group cohesiveness and self-consciousness. Though deeply aware of their shared status as *Handwerker*, the artisans derived their most fundamental sense of identity from their membership in a specific craft. This sense of group identity, it should be emphasized, did not arise simply from belonging to a specific branch of production. Shoemakers did not feel a common bond with all other shoemakers; indeed, some shoemakers might regard other practitioners of the same craft as their most bitter enemies. The common bond was shared by those members of any given craft who adhered to the rules of the craft as articulated and enforced by formal organizations of the *Handwerk* which crossed city and territorial lines.[29] At times these organizations reviewed and renewed their policies at regular regional meetings. But even when this was not the case, solidarity was maintained and behaviour constrained by a system of inter-urban communication maintained largely by journeymen as they travelled from one city to the next. This solidarity with a broader network of fellow artisans could give members of a specific guild some leverage in their dealings with the local authorities. No matter how firmly it might assert its ultimate right to issue, revise or confirm the statutes of every *Handwerk* in the city, no council would want to impose conditions that could lead to local artisans being boycotted or blacklisted by other members of their craft.

The political role of artisans in the German city is often difficult to gauge because most of the time relations between the specific guilds and the municipal authorities were non-confrontational. Lines of authority which were clear in theory could easily be blurred in practice. In Augsburg, for example, the political and financial autonomy of the guilds had been drastically reduced in 1548. In principle each craft was placed under the direct control of *Vorgeher* – members of the craft chosen by the authorities to supervise their colleagues. But in fact the role of the *Vorgeher* was somewhat ambiguous. In his masterly study of Augsburg's baking trade in the early seventeenth century, Bernd Roeck assesses the role of the bakers' *Vorgeher* in a way which, he suggests, may have applied to many other crafts as well:

> In the eyes of the council the *Vorgeher* were not so much representatives of the craft as functionaries who played an integrating role yet also supervised their own colleagues. In fact they had two functions at once: they were 'buffers' between the craft and the council, who represented ... the interests of the craft just as energetically as they must have worked within their organization for obedience to the terms of the [municipal] craft ordinance ... No matter how loyal they felt to the authorities, in their own self-perception - insofar as this can be determined from the sources about their everyday activities - they saw themselves more as representatives of the interests of the craft than as agents of the council.[30]

Interest-group politics in the early modern city was normally conducted in the idiom of the time, and that idiom usually embodied the articulation of grievances or submission of respectful petitions which the authorities would be expected to address with paternalistic concern. In many cases, of course, the magistrates had to adjudicate between different groups with competing interests - between merchants who wanted less economic regulation and artisans who wanted more, between artisans from different crafts who found themselves in competition with each other, or even between richer and poorer members of the same craft who had found it impossible to resolve their differences internally. Whatever the magistrates did, some groups would inevitably be more satisfied than others, but most of the time municipal authorities were able to balance out the claims of different groups in ways that preserved stability and social peace.

Under these conditions of routine interest-group politics, the potential of the artisans for collective political action was not always apparent. Much of the time, in fact, groups of artisans functioned politically in isolation from one another. Considerable political energy went into persuading the magistrates to accord some privilege or grant some concession to the masters of a particular craft, or even - as in the case of the large and economically heterogeneous weavers' guild of Augsburg - to a particular sub-group within the craft.[31] But there were also times when the artisans pursued their aims more aggressively and even, if conditions warranted, reached far beyond the concerns of a single craft to operate as a collective group.

This could happen, for example, under conditions of political crisis, for occasionally a city's political equilibrium would collapse. When a sufficiently large sector of the citizenry concluded that the magistrates were not acting on behalf of the common good, traditional forms of petitionary politics would give way to an open confrontation. The leadership in these struggles of the *Bürgerschaft* against the *Rat* normally came not from the *artisanat* but from social groups ranked just below the municipal elite – men who, in many cases, felt that the existing council had blocked their legitimate aspirations for a greater voice in civic affairs. Artisans nevertheless often played a pivotal role in conflicts of this sort.

A classic study of the cycle of disturbances in Aachen in the late sixteenth and early seventeenth centuries emphasizes that the lines of conflict were drawn primarily between different sectors of the civic elite: 'The broader strata of artisans', Heinz Schilling reports, 'functioned only as participants in mass actions, which is to be understood in connection with their economic dependence on the leaders of the revolt.'[32] Yet valid as this analysis may be for the case of Aachen, it would be a mistake to assume that artisans were always relegated to such a dependent role in civic conflicts. It is certainly true that artisans hardly ever initiated a struggle against the magistrates, but once such a conflict erupted the artisans often recognized opportunities to pursue their own interests more aggressively than was possible in the conditions of normal municipal politics.

This, as we have seen, was exactly what happened in Reval in the 1670s, but the same strategy was apparent in many other episodes of urban conflict, including the most famous of all urban disturbances in seventeenth-century Germany – the Fettmilch Uprising of 1612-14 in Frankfurt am Main.[33] The uprising began as a two-pronged attack by a broad coalition of disaffected citizens, both against the corrupt policies of the city council and against the economic privileges of the city's Jewish population. Individual artisans were always among the most ardent supporters of the oppositional movement. But, in addition, one craft guild after another made sure to get a statement of its own economic grievances included among the briefs submitted to the imperial commissioners who were sent to investigate the conflict.[34] For the citizens' uprising opened up a sudden opportunity – which almost every guild was swift to seize – to pursue traditional aims in what seemed to be a more promising political framework. Even more generally, the leaders of the oppositional movement demanded that any citizen who did not yet belong to a guild or comparable society be forced to join one. The extent of disaffection among the citizenry impressed the imperial commissioners, and many of the citizens' demands were specifically granted in the great compact of December 1612 which was negotiated in order to resolve the conflict.[35] In fact the conflict resumed and the objectives of guild members were eventually forgotten as the uprising escalated into alarming outbursts of anti-patrician and anti-semitic violence. When the rebellion was finally crushed and its leaders executed, the artisans of Frankfurt were, if anything, worse off than before. But this should not obscure the way in which they attempted to use the new conditions of a civic conflict to pursue their traditional objectives.

It was only to be expected that even in times of political upheaval, as in more peaceful times, the artisans' perception of their interests should still be framed largely in terms of their particular crafts. The guild, after all, was the most important single associational grouping to which any artisan belonged, and the guilds themselves occasionally fell into jurisdictional disputes with each other. Yet the identification with particular guilds did not eliminate the artisans' capacity to look beyond the needs of a single trade so as to act in solidarity with all other craft masters.

The struggle between the council and citizens of Lübeck in the 1660s provides a striking example of this.[36] In 1661 the patrician-dominated city council, facing an insurmountable financial crisis, reluctantly asked representatives of different segments of the citizenry to support increases in taxation. The city's seven merchant guilds responded that they would do so only if citizens were given more control over the city's financial affairs. When it became clear that the council would resist such a constitutional change, the merchant guilds openly solicited the support of the city's craft guilds, whose leaders responded that they would only join the movement if their own objectives were adopted. As in many cities, the artisans were eager to diminish competition from unguilded craftsmen who worked in the countryside. What gave the situation in Lübeck an unusual twist is that many of the hated rural artisans lived and worked on country estates owned by the city's own patrician magistrates. Not surprisingly, the council stalled in meeting the citizens' various demands, but this only heightened the artisans' solidarity. In the spring of 1665, a grand army of hundreds of Lübeck artisans marched into the countryside to attack the rural craftsmen on the patricians' estates: brewing vats were overturned, looms were smashed, tools destroyed. Alarmed by these developments, the magistrates soon agreed to a new system of financial administration; some of the most hated council members resigned and left the city. Constitutional reforms, which were ratified by an imperial commission in 1669, significantly reduced the powers of the city council. The intervention of the city's artisans at a crucial moment had certainly played a major role in bringing these changes about.

Reval, then, was far from unique. To be sure, artisans were never the strongest players in the politics of the seventeenth-century German city – but they were rarely a negligible factor. The artisans did not always constitute a unitary 'interest group': each craft had its own specific interests, and even within any craft there could be significant tensions between richer and poorer masters, between masters and journeymen, or even within households along gender or generational lines. Yet most artisan masters did have a clear perception of group purpose and a powerful sense of group identity, both as masters of their own crafts and as craft masters in general. And this is what made it possible for them to function with considerable effectiveness in the interest-group politics of the seventeenth-century German city.

Notes

1. A population of 9801 is reliably recorded for 1708: Stefan Hartmann, *Reval im Nordischen Krieg*, Bonn-Godesberg: Verlag Wissenschaftliches Archiv (1973), pp. 69-70. Estimates for the seventeenth century are provided by Ernst Gierlich, *Reval 1621 bis 1645: Von der Eroberung Livlands durch Gustav Adolf bis zum Frieden von Brömsebro*, Bonn: Kulturstiftung der deutschen Vertriebenen (1991), pp. 279-81, and Johann Dietrich von Pezold, *Reval 1670-1687: Rat, Gilden und schwedische Stadtherrschaft*, Cologne/Vienna: Böhlau (1975), pp. 10-11.
2. Cf. Arnold Soom, *Der Handel Revals im siebzehnten Jahrhundert*, Wiesbaden: Otto Harrassowitz (1969).
3. For an overview of Reval's political structure, see Gierlich, *Reval*, pp. 21-57.
4. Cf. Paul Johansen and Heinz Von zur Mühlen, *Deutsch und Undeutsch im mittelalterlichen und frühneuzeitlichen Reval*, Cologne/Vienna: Böhlau (1973), pp. 65-6, 123-6, 286-94.
5. Arnold Soom, *Die Zunfthandwerker in Reval im siebzehnten Jahrhundert*, Stockholm: Almqvist & Wiksell (1971), pp. 120-74; Pezold, *Reval*, pp. 110-29, 367-9.
6. Soom, *Zunfthandwerker*, pp. 138-9.
7. Ibid., p. 145.
8. Ibid., pp. 123-4; Gierlich, *Reval*, pp. 57-8.
9. Soom, *Zunfthandwerker*, pp. 149-67.
10. Pezold, *Reval*, 110-35.
11. Soom, *Zunfthandwerker*, pp. 81-4; Pezold, *Reval*, pp. 130-75, 259-67, 324-7.
12. For a recent survey of the literature, see Wilfried Reininghaus, *Gewerbe in der frühen Neuzeit*, Munich: Oldenbourg (1990).
13. Edward C. Banfield and James Q. Wilson, *City Politics*, Cambridge, MA: Harvard University Press (1963), p. 7.
14. Cf. Murray S. Stedman Jr, *Urban Politics*, 2nd ed., Cambridge, MA: Winthrop Publishers (1975), pp. 209-20.
15. This predominantly 'elitist' approach is apparent, for example, in the otherwise excellent survey by Klaus Gerteis, *Die deutschen Städte in der frühen Neuzeit: Zur Vorgeschichte der 'bürgerlichen Welt'*, Darmstadt: Wissenschaftliche Buchgesellschaft (1986), pp. 65-97, where urban politics is seen predominantly in terms of constitutional norms and municipal administration.
16. An influential expression of this argument was provided by Bernd Moeller,

Imperial Cities and the Reformation, trans. by H.C. Erik Midelfort and Mark U. Edwards Jr, Philadelphia: Fortress Press (1972), esp. pp. 103-15.
17. For extensive references to the literature on this subject, see Christopher R. Friedrichs, 'German Town Revolts and the Seventeenth-Century Crisis', *Renaissance and Modern Studies*, 26, 1982, pp. 27-51, and idem, 'Urban Conflicts and the Imperial Constitution in Seventeenth-Century Germany', *Journal of Modern History*, 58 (supplement), 1986, pp. S98-S123.
18. See Eberhard Naujoks (ed.), *Kaiser Karl V. und die Zunftverfassung: Ausgewählte Aktenstücke zu den Verfassungsveränderungen in den oberdeutschen Reichsstädten, 1547-1556*, Stuttgart: W. Kohlhammer Verlag (1985).
19. Siegfried Müller, 'Kontinuität und Wandel innerhalb der politischen Elite Hannovers im 17. Jahrhundert', in Kersten Krüger (ed.), *Europäische Städte im Zeitalter des Barock: Gestalt - Kultur - Sozialgefüge*, Cologne/Vienna: Böhlau (1988), pp. 223-69.
20. Cf. Heinz Schilling, 'Wandlungs- und Differenzierungsprozesse innerhalb der bürgerlichen Oberschichten West- und Nordwestdeutschlands im 16. und 17. Jahrhundert', in Marian Biskup and Klaus Zernack (eds), *Schichtung und Entwicklung der Gesellschaft in Polen und Deutschland im 16. und 17. Jahrhundert: Parallelen, Verknüpfungen, Vergleiche*, Wiesbaden: Franz Steiner Verlag (1983), pp. 121-73.
21. Cf. Bernd Dreher, *Vor 300 Jahren - Nikolaus Gülich*, Cologne: Kölnisches Stadtmuseum (1986), pp. 10-20.
22. Mack Walker, *German Home Towns: Community, State and General Estate, 1648-1871*, Ithaca: Cornell University Press (1971), esp. pp. 34-72.
23. Cf. the classic article by Otto Brunner, 'Souveränitätsproblem und Sozialstruktur in den deutschen Reichsstädten der früheren Neuzeit', *Vierteljahrschrift für Sozial- und Wirtschaftsgeschichte*, 50, 1963, pp. 329-60.
24. Cf. Christopher R. Friedrichs, 'Urban Politics and Urban Social Structure in Seventeenth-Century Germany', *European History Quarterly*, 22, 1992, pp. 187-216, esp. pp. 194-5.
25. Cf. Olaf Morke, 'Der gewollte Weg in Richtung "Untertan": Ökonomische und politische Eliten in Braunschweig, Lüneburg und Göttingen vom 15. bis ins 17. Jahrhundert', in Heinz Schilling and Herman Diederiks (eds), *Bürgerliche Eliten in den Niederlanden und in Nordwestdeutschland: Studien zur Sozialgeschichte des europäischen Bürgertums im Mittelalter und in der Neuzeit*, Cologne/Vienna: Böhlau (1985), pp. 111-33.
26. Pezold, *Reval*, p. 287.
27. Cf. Friedrichs, 'Urban Conflicts'.
28. The number of professional soldiers employed by the council in the 1660s is not known, but in 1622 the total number of Stadtsoldaten in Reval came to about twenty: Gierlich, *Reval*, p. 273.
29. Cf. Frank Göttmann, *Handwerk und Bündnispolitik: Die Handwerkerbünde am Mittelrhein vom 14. bis zum 17. Jahrhundert*, Wiesbaden: F. Steiner Verlag (1977).
30. Bernd Roeck, *Bäcker, Brot und Getreide in Augsburg: Zur Geschichte des Bäckerhandwerks und zur Versorgungspolitik der Reichsstadt im Zeitalter des Dreissigjährigen Krieges*, Sigmaringen: Jan Thorbecke Verlag (1987), p. 160.

31. Cf. Claus-Peter Clasen, *Die Augsburger Weber: Leistungen und Krisen des Textilgewerbes um 1600*, Augsburg: Verlag Hieronymus Mühlberger (1981), esp. pp. 140-210, 237-60.
32. Heinz Schilling, 'Bürgerkämpfe in Aachen zu Beginn des 17. Jahrhunderts: Konflikte im Rahmen der alteuropäischen Stadtgesellschaft oder im Umkreis der frühbürgerlichen Revolution?', *Zeitschrift für historische Forschung*, 1 (1974), pp. 175-231, here p. 207.
33. The vast literature on this event is discussed by Christopher R. Friedrichs, 'Politics or Pogrom? The Fettmilch Uprising in German and Jewish History', *Central European History*, 19, 1986, pp. 186-228.
34. As in the 'Gravamina in puncto illicitarum exactionum' (Sept./Oct. 1612), text in Friedrich Bothe, *Frankfurts wirtschaftlich-soziale Entwicklung vor dem Dreissigjährigen Kriege und der Fettmilchaufstand (1612-1616)*, vol. 2, Frankfurt am Main: Joseph Baer & Co. (1920), pp. 421-48.
35. Text ibid., pp. 492-510.
36. Jürgen Asch, *Rat und Bürgerschaft in Lübeck, 1598-1669: Die verfassungsrechtlichen Auseinandersetzungen im 17. Jahrhundert und ihre sozialen Hintergründe*, Lübeck: Verlag Max Schmidt-Römhild (1961), pp. 99-173.

CHAPTER THREE

Cultural analysis and early modern artisans

James R. Farr

What was an artisan?[1] A deceptively simple question becomes surprisingly complex when we shift away from the traditional frameworks in which this question could be answered towards one informed by cultural analysis.[2] One could respond to this question as many historians have before, that artisans were members of guilds, or, if such an institutional framework was found unsatisfactory, one could offer a production-centred definition, that artisans were skilled people who fashioned artefacts with their hands and tools but without the aid of machinery, the classic handicraftsman. Both approaches have lengthy historiographical pedigrees, and elements of truth in them.[3] But recently, some historians have found these traditional frameworks inadequate to analyse important aspects of the experience of the groups of people – men and women – we have labelled 'artisans'. Not every such person, in fact, belonged to a guild (virtually no women did), nor were weavers (as they would be the first to tell us) simply men or women who happened to weave thread, bakers simply men or women who happened to cook bread. To grasp the sense that these men and women had of themselves, and that others had of them, requires moving beyond an institutional or productive (even economic) framework towards one that can accommodate meaning in general and questions of identity in particular.

We need to explain how artisans constituted their identities, accounting for the 'insider's' view and activity, in Peter Burke's words, as well as those of the 'outsiders';[4] better, we need to demonstrate the relationship between these perspectives and actions. I suggest in what follows that artisans living in European cities between the fourteenth and, in some places, the late nineteenth centuries were defined not primarily as producers, as their labels may suggest, but rather as an *état*, a rank or 'degree', a *stand*, and, moreover, they thought of themselves as such. Such social and self-definitions were rooted in a cultural experience which included but also transcended production; these definitions were profoundly influenced by shifts in political, legal, intellectual, as well as economic developments across these centuries. Artisans did not make themselves in isolation, nor were they hapless victims simply moulded by forces beyond their control. They were products of their own ceaseless struggle, not just to survive, but to maintain rank and a sense of social place in the face of powerful, often inimical forces in their world, turning these forces to their advantage when they could, suffering fragmentation or transformation when they could not.

Let's begin with labels: 'artisan', 'tailor', 'shoemaker' all refer to a function, indeed, a seemingly economic one. But before we functionally tie such labels to production and deduce historical meaning directly from that association, we should consider that such labels are 'signs' defined in particular historical and cultural settings.[5] 'Rethinking' the meanings of labels has been recently pursued by scholars of eighteenth- and nineteenth-century labour history - notably Simona Cerutti, Joan W. Scott, William Sewell, Michael Sonenscher, Donald Reid and William Reddy - and some of their ideas have conveniently been presented in a single volume.[6] 'Rethinking labour history' has entailed revising a number of fundamental issues that were once taken for granted, but which now seem increasingly open to question. The role of capitalism as a motor of social change, the proletarianization of labour, the formation and meaning of class and, therefore, 'class consciousness' - all have been questioned by historians working from post-structural assumptions which challenge the notion of self-evident 'reality' which many investigators had simply assumed. Inspired by interpretive anthropology, linguistics, and some literary theory, revisionist labour historians are calling for 'a fundamental overhaul of the field'.[7]

My goal here is more modest, but I would like to suggest how a cultural approach to labour history might fruitfully be extended to an analysis of early modern artisans. Elsewhere I am taking this approach to a synthesis of the entire spectrum of the European artisan experience from the fourteenth to nineteenth centuries,[8] but here I shall concentrate on two related topics of considerable interest to labour historians recently: journeymen brotherhoods, and the meaning of skill. I shall attempt to demonstrate the merit of embedding in cultural contexts ostensibly economic subjects like skill or worker organization. Throughout I shall emphasize status rather than economics or production, and I shall invoke the heuristic concepts of identity, difference and distinction. In the historical context of the hierarchical world of early modern Europe, identity (artisanal, or any other) was constructed through erecting and maintaining boundaries between an imagined 'us' and 'them', and so identity was rooted in, as Peter Sahlins puts it, 'a subjective experience of difference'. It was, therefore, relational, and contingent upon context.[9]

Ancien régime taxonomy was a structured system of hierarchical differences, a structure which was nonetheless dynamic, fragile and unstable. It was within this structure, a product of an incessant interrelationship of prescription and practice, that individuals and groups of individuals purposefully engaged in the constitution of meaning in their lives. They did this, I shall suggest, through reflexive relationships that were continually mediated by symbolic exchange with and within an 'always already' culture.

Culture and the problem of labour

Marx, as is well known, isolated labour as the human self-generative principle, and assumed that economic rationality was essential to the labour process which followed directly from the natural necessity of production. In other words, Marx, for all his historicism, nonetheless 'naturalized' labour no less than the classical economists like Adam Smith or David Ricardo before him, making it the foundation of the edifice of culture.[10] Most historians of artisans, Marxist or not, have similarly 'essentialized' labour, assuming a direct functional relationship between this activity and what an artisan was. The merit of a cultural approach to labour is that, by embedding economic process in cultural contexts, it eschews a reductive economism.[11] To avoid ahistorical reification or abstraction that so often accompanies economism, a more analytically sufficient approach would be one that holds cultures to be 'always already' at the moment of analysis. To assume otherwise leads the historian to 'essentialism', to privileging a particular, anterior element within the cultural matrix as the fundamental one.[12]

Cultural history can profitably start from a premise that, in Marshall Sahlins's words, 'cultures are meaningful orders of persons and things'.[13] Such a premise is undergirded by the assumption that experiences (even economic ones) can have no intrinsic or natural meaning; experience, even labour, only gains meaning by symbolic inscription in a relational cultural system.[14] Labour and production, then, should not be thought of as prior to or outside culture, but already embedded in it at the moment of analysis and as cultural as any other human activity (and therefore meaningless outside the cultural frame).[15] Productive labour was only one constituent part of a multidimensional artisan identity and experience. Even production, as Patrick Joyce points out, 'is inexplicable without an understanding of reproduction, both in its limited sense as the reproduction of labour power and in its larger sense as the reproduction of society itself ... The political and social are [thus] inseparable from the economic'.[16] Indeed, in contemporary capitalistic societies, theorists have argued that work and community are interpenetrating spheres, and thus work and non-work cannot be rigorously separated analytically without distorting the meaning of both.[17] Artisans, after all, were not concerned with nothing but their work – indeed, they may have been indifferent to it, or even despised it[18] – and so a cultural analysis of artisans seeking to understand identity and social classification (and the relationships between them) can do more than view artisan culture as based on the techniques of work.

If labour should not be privileged as the defining human activity and the bedrock of culture, production should not be taken as a determinant of identity, and artisans should not be essentially defined by the kind of labour they perform, then how should we isolate this unit of analysis? To return to our original question, what was an artisan? And how do we study him or her historically? Let's consider a theory of cultural distinction. Bourdieu perceives a cognitive operation in human agents which classifies their perceptions and their practical activity in an analysis relational to their position within the social whole. Fundamentally, such positions are constituted within systems of differences communicated symbolically.[19] That is, identity is formed by recognition of similarity and difference, and social groups – what William Reddy calls 'bundles of association' – are formed by the related and often contestatory process of the recognition of inclusion and exclusion.[20] The resulting social group coheres through a sometimes explicit, but often tacit, code which enables a coherence of a specific kind. Individuals can and do belong to multiple groups, resulting often but not necessarily in a hierarchical valuation of the various groups by the individuals so engaged. In any case, symbolic exchanges within the group and structured within the code enable 'the possibility of apprehending others, their social condition, and thereby their relation to oneself "at first glance"'.[21]

A great deal of empirical evidence commends this theory. Rather than finding his or her social being defined by his or her labour, then, an artisan lived his or her life (and his or her work) as a meaningful relation of symbolic exchanges, where labouring activity was a sign of social place as much as a means to survival or material accumulation. Such exchanges were ephemeral encounters in contingent situations, and so were simultaneously dynamic and structured by a shared system of communication in which meaning inhered.[22] Because demographics and social mobility rendered friendships fleeting and social groups fragile, networks and alliances were continually in need of re-creation and reconstitution. It was through these encounters and exchanges of 'symbolic capital' that artisans continually constructed and reconstructed their sense of a coherent identity, remembering from the immediate past the attributes that defined them while plunging ineluctably forward into a context forever in flux. Simultaneously and inextricably they established and re-established their place within the taxonomic structure of society through apprehension of difference, distinction and status.[23]

The resulting taxonomy has apparent structural qualities, but it is nonetheless unstable because symbolic exchange paradoxically destabilizes hierarchy while it demonstrates it. Symbolic exchange is necessary to construct the order of distinction, to visualize it so it can be 'read', but such exchange is simultaneously rooted in a process of endless contestation and negotiation and, as a result, like all exchange, is indeterminate and resists closure. The concept of symbolic exchange is thus fundamentally historical.

In sum, it seems to me that a cultural analysis of early modern artisans – understanding their identity as well as their place within the social structure – might proceed in part across a series of interconnected axes of social differentiation and association. In the rest of this essay I want to explore the concept of status within these methodological parameters. In this way my unit of analysis, the artisanate, can be isolated and defined relationally without privileging any particular aspect of their existence (their work life, or their place in productive relations, for example). Such an analytical premise of social differentiation may be especially apt for study of the early modern age for, as many historians have recently shown, the theme of social distinction and differentiation increasingly preoccupied the men and women of this period.[24]

Distinction and the artisanate

Journeymen brotherhoods

In the fourteenth century everywhere in Europe stratification and differentiation were becoming increasingly articulated, reaching their apogee in the seventeenth century and continuing into the eighteenth. Bernard Chevalier has persuasively contended that in the fourteenth century, in the 'good towns' of France, an internal hierarchy emerged which loosely divided the urban populace into three vertically ranked categories, granting diminishing degrees of 'honour' to each rank, from *les bons* (the bourgeois), to *les communs* (craftsmen of guilds), to *les vulgaires* (labourers outside the guild).[25] This loose structure heralded the *mise en place* of the corporate regime in which craft guilds became defined constitutionally as the *corps de métiers*. Craft guilds, of course, predated the corporate regime,[26] but only in the fourteenth century do we find them specifically organized as *corps* and correspondingly assuming an *état*, a normative quality that placed them hierarchically in the social and constitutional firmament of the polity. The *gens de bras*, or workers not organized in *corps*, had no *état*, and thus were designated *gens sans qualité*.[27]

From the outset, then, corporatism was a new, prescriptive system of order with economic dimensions, but one that was also inextricably linked to hierarchy and distinction as well as politics. It was grounded in a demand for the subordination and discipline of inferiors. Indeed, the corporate regime gained definition by the principle of exclusion. Workers without *état* were defined outside the system of order, and consigned as a result to the netherworld of disorder. These men and women were imagined by those with a stake in the new system to 'have in their head only the inversion of all values, the abolition of providential differences'.[28] Thus, when workers did challenge their masters over control of the labour market, for example, their actions were interpreted by masters, no less than municipal and later royal authorities, as not simply economic disturbances but as violence against order *per se*, and were invested with cosmological, not just narrowly economic, importance.

Throughout Europe, as corporatism and the *corps de métiers* took hold, journeymen began forming associations themselves. In addition to playing an important role in the struggle with masters over the labour market, these brotherhoods became the crucible of identity construction for journeymen as these artisans rigorously excluded the *gens de bras* and women from their communities. Journeymen brotherhoods became increasingly well organized across the early modern centuries.[29] True, their networks served as labour clearing-houses, distributing workers among the shops of a given town and even among the towns of the various kingdoms, but these brotherhoods, no less than the corporations, also explicitly defined *état*, degree, or *stand*, even if the corporate regime did not officially acknowledge it. Indeed, journeymen styled themselves as a 'mobile rank', an '*état sans domicile*'.[30]

Journeymen everywhere, no less than their masters, were preoccupied with their honour,[31] insults to which, surprisingly, were probably as common a cause for strikes and boycotts as disputes over wages or working conditions.[32] These proud men were extremely sensitive to their *honnêteté* and were ever ready to bolt from a shop or town if it was besmirched. Honour, after all, was a possession that signalled social place and respectability, and for journeymen it was what distinguished them from the wage-workers, beggars or vagabonds – from all of those people 'without quality' who were so disconcertingly similar to the journeymen in their mobility.

A self-styled sense of honour also distinguished journey*men* from women workers. As journeymen associations formed in the late middle ages, they confined their membership exclusively to males and, in the early modern period, they became aggressively anti-female, refusing to work in shops of masters where women were employed.[33] In 1649, for instance, journeymen boycotted the master hatmakers of Frankfurt, refusing to work in the same shops as journeymen who had dishonoured themselves by working alongside women in hatter workshops in Fulda. Such exclusionary tactics reduced the number of workplaces available to journeymen and so ran counter to their immediate economic interests, but the journeyman's honour, and thus his status, was paramount.[34]

Among the symbols of his status were the journeyman's tools, which he owned and which were more than a means to earn a living; they signified the possession of a personal qualification. By the eighteenth century, at least in France, this self-consciousness of distinction and *état* was overlain by ideas about liberty. Journeymen boycotted and struck not just to demand higher wages, piece rates, or control over worker placement, but more often to proclaim their right to exist, as Arlette Farge puts it, 'in a free relationship' with their masters which meant 'the right not to be considered a servant', and, as some journeymen concluded, entitled them to carry a sword.[35]

Robert Darnton has described in exceptionally lucid detail how eighteenth-century journeymen in the printing trade elaborately constructed their exclusive community through ceremony, ritual and distinction that clearly defined internal rank and set them off in their estimation in society as *états*, as they explicitly put it.[36] Hans-Ulrich Thamer and Merry Wiesner find evidence strikingly similar to Darnton's in early modern Germany, where ceremonies like strictly observed greeting formulas (the *Schenke*) and gift-giving crisply distinguished insiders from outsiders, and regulated the passage of the latter into the former.[37] As the eighteenth-century French glazier Jacques-Louis Ménétra reveals so well in his journal, a journeyman's identity was tightly integrated into these groups.[38] Institutions like these, with symbolic gatekeeping, treasuries, officers, houses of call and rolls of members, have been found in France, Germany and England at least since the fifteenth century, and everywhere exclusiveness separated them from another class of worker, the 'unqualified'.[39] The journeymen's impulse to define rank was no less pronounced than that of the masters, even if it had to be done in the shadow rather than the full legal light of corporatism.

Cynthia Truant draws the most complete picture we have of these brotherhoods in France, institutions which gained structure as journeymen were increasingly excluded from the masters' guilds. In these shadowy institutions she finds that journeymen redirected Catholic and corporate ritual to their own ends, with ceremonies akin to baptism (replete with renaming and initiation into the new community), communion (sealed by libations of a rather secular kind, quaffed in a tavern), and godparenthood providing the sinews of community. Within their ranks, the boundaries of which were so self-consciously constructed and visually demonstrated by ritualized welcoming (the *bienvenue*) and leave-taking (the *conduite*) of brothers on the *tour de France*, they elaborated hierarchies of their own, again paralleling developments in the culture at large. Captains and lieutenants saw to it that discipline was enforced, and increasingly in the late seventeenth and eighteenth centuries this discipline took on a 'civilizing' function, with the brotherhoods' leaders insisting on *politesse* and self-control, most notably in the insistence to moderate drinking, to cease brawling, and to confine indelicate language to its 'appropriate' venue.[40]

Such communities, of course, deeply alarmed the masters as well as the political authorities. Their consternation, however, stemmed not just from the feared economic ramifications of worker control of labour markets, wage levels, or piece rates. More fundamentally, masters and magistrates sensed a challenge to discipline in general. Such a 'crisis of discipline' threatened something more elemental than the financial security of the master's business, as important as that may have been; it undermined the hierarchical structure of society and the master's, and ultimately the magistrate's, place within it. Everything rested on the idea of subordination, and insubordinate workers – tellingly described by masters in language redolent of hierarchy as 'crude and vile animals' – signalled a challenge to order that reverberated on mundane and cosmic levels.[41] It raised the spectre of the masterless man, a vision that haunted the minds of master guildsmen no less than the political authorities of every *ancien régime* polity. By threatening to step outside the sanctioned hierarchy, masterless men exposed the fissures between the legitimate, prescriptive taxonomy on the one hand, and social existence on the other. As such, the challenge struck at the very base of the master's status and identity.[42]

The meaning of skill

Until recently, histories of artisans often described them as pre-industrial skilled workers destined to be crushed by the juggernaut of mechanized, factory production. In the process they found their skills eroded. Few historians would deny that some artisans became proletarianized, deskilled factory workers, but recent research also amply demonstrates that in some sectors of the economy workers were reskilled to suit the needs of mechanized production, while a great many others remained skilled artisans working in the traditional way. Moreover, many of these artisans, still practising their trades in small workshops, were important, if increasingly vulnerable, participants in industrialization into the mid-nineteenth century in some places, and the late-nineteenth century in others.⁴³

Within this historiography, some historians have also come to rethink the meaning of skill itself. Should skill be objectively defined as a function of production?⁴⁴ John Rule has written that skill to eighteenth- and early nineteenth-century English artisans was a property and thus a 'distinguishing mark separating the artisan from the common labourer'. As much as technical aptitude in the fashioning of products, skill was 'symbolic capital', the possession of which 'entitled its holder to dignity and respect'. It was conferred by a community and protected as a 'right' by tradition and custom.⁴⁵ Michael Sonenscher similarly emphasizes that skill had a great deal to do with difference and distinction. Focusing on the world of journeymen, he points out that prowess to do a job well was certainly valued and recognized among these men, but such skill served a social purpose as well as an economic one. In a world where a vast pool of labourers worked in trades with a 'relatively limited range of materials' and that saw few 'rigid technical and occupational divisions', there was a wide dissemination of similar abilities. So when journeymen brotherhoods ritually demonstrated the feats of prowess of their members in competitions, they were securing their community and articulating its boundaries by transforming 'similarity into difference in a world in which too many people could do the same thing'.⁴⁶

Recent research on women's work has deepened our extra-economic understanding of skill, too. Female 'skills' had been undergoing a process of devaluation for centuries before industrialization, a process which had nothing to do with technology, factories or fixed capital.⁴⁷ Instead, since the devaluation of female skills was inversely related to the definition of skill by men, such 'deskilling' was a product of particular social and gender relations. Among the lessons we learn from this gloomy process is that it is not enough to define skill as manual facility, as 'the co-ordination of perceptual and motor activity', although this certainly is an important part of it. Nor is it sufficient to define skill as knowledge of the properties of materials and how to assemble them into products, however necessary that is. Such a definition, useful as it is, is overly production-centred and, moreover, ahistorical. Seamstresses, after all, according to this definition, were skilled, yet they were construed by the men of their world to be 'unskilled'. Production-centred definitions tell us nothing, in fact, of the *meaning* of skill to those who claimed to have it, and to those who wrote about it. To understand the meaning of skill, we must situate it contextually, for it is a relational quality, measured against those who supposedly possessed less of it, or none at all – the semi-skilled and the unskilled.

Two historians – Stephen Marglin and Harry Braverman – have

suggested that what the industrial revolution was really about was not the economic efficiency that came from the factory system, the subdivision of the division of labour, or scientific management, but rather the social power that capitalists achieved from greater control of the workforce and the workplace, effected through hierarchy and the discipline of labour.[48] This may have been true for the industrial period, but it is a story with a long history, since this had been a fundamental concern of master guildsmen since the fourteenth century. Hierarchy and the discipline of labour were principles enshrined in guild ordinances, and these were to be effected through, in part, control over skill – both its meaning and its possession.

Artisans as a group of guildsmen or, more rarely, women, were distinguished from other groups and individuals in this hierarchical society legally and constitutionally by corporatism, the legal armature of every polity in Europe until the late eighteenth century in some places, the nineteenth in others. As a normative means of social classification and social control, corporatism designated artisans by trades. Occupational nomenclature, though not always a sure guide to the kind of work performed,[49] more importantly to early modern Europeans, *was* a sure designation of the social attributes of the individual, his or her *qualité*. As Steven Kaplan correctly observes, the organizing principles of the corporate idiom were embedded in a social taxonomy, but this in turn was closely linked to the exercise of power: 'the tools of distinction used to forge the classification system are tools of social and political control'.[50] Such a hierarchical system was, then, equally a power structure, and distinction and difference were animated by a concern for subordination and discipline of inferiors, be they journeymen, apprentices, or women. Breach of discipline by journeymen or wage-workers reflected more than instability in the labour market; also, and more dramatically, it reflected a perceived threat to hierarchy and the principle of distinction itself. Masters were deeply sensitive to insubordination by journeymen and wage-workers, and journeymen were keen, in turn, on maintaining the inferiority of wage-workers beneath them.

The meaning of skill for masters became increasingly associated with discipline and subordination. Apprenticeship as an institution was designed to serve these ends, as was legislation demanded by guildsmen against 'illicit' clandestine workers. Apprenticeship regulations were enshrined in most guild statutes from the fourteenth and fifteenth centuries, and these hand-picked boys (not every worker in the shops served an apprenticeship) tracked for mastership were evaluated at the end of their terms. Their skills were assessed but, more importantly, they were judged whether as a person they were, as the French put it, *idoine et souffisant* ('appropriate and satisfactory'). The hatters of Dijon stated this explicitly in their statutes drawn up in 1487. One clause dictated 'that the one who will wish to become a master of the said trade [must] be found *appropriate and satisfactory* by the aldermen of the town and the officers of the guild'. Just what 'appropriate and satisfactory' meant is never explained.

Personal quality and technical skill were, in the minds of masters, inseparable, and the criteria for either are never explicitly laid down. Take the statutes of the Dijon goldsmiths codified in 1443 as an example. In a trade where technical aptitude must have been at a premium, in the clause dealing with apprentices, this is all the statutes say: 'That the sons of masters of the said trade and the apprentices who will have completed their term of six years will be able to open a shop in the said town, [after] the children of the masters host a dinner for the officers of the guild ... and the apprentices pay 60 sous ... .'[51]

In some trades admission to mastership entailed the making of a 'good and satisfactory [*bon et souffisant*] masterpiece', the quality of which would be evaluated by the aldermen and the officers of the guild. Just what criterion was invoked, however, the statutes never clearly say. The by-laws of the Dijon hatters are as close as we come:

> The one who wishes to become master of the trade will make at his own expense his masterpiece of three hats ... in the following manner: a black hat *à la verge frisé* made of good, fine wool and which will be judged by the officers of the guild and the aldermen [of the town whether] it is good and sufficient [*bon et souffisant*]; a white hat *crêpé à l'esguille* also of fine wool and which will be judged by the officers and aldermen ... as good and sufficient [or not]; and a coarse black hat *à la verge* cut from good but coarse wool ... and will be judged by the officers and the aldermen as good and sufficient [or not].[52]

Clearly apprentices were trained in the skills of the trade, but apprenticeship was about much more than the acquisition of technical aptitude. Non-apprenticed 'skilled' workers received that, too. Rather, apprenticeship was a special education for carefully selected hopeful future masters that instilled in them the values of their future community. Discipline and subordination were chief among them, as apprenticeship contracts and indentures make abundantly clear. Apprentices lived and worked in their masters' homes and workshops as if they were the masters' children, and as such were under the strict thumb of a surrogate father whose authority was sanctioned by a patriarchy that informed the entire power structure of all early modern polities. Contracts between parents and masters over the apprenticing of the son ideally secured 'the advantage' such a position held for the boy. In return for a fee, the apprentice was, as Arlette Farge puts it, 'fed, warmed, washed, lodged, and lit'. He was instructed in the craft 'without [the master] concealing anything', but, as importantly, he was to learn 'good manners and healthy conduct' as well. The apprentice, for his part, was bound to his new master and was required to obey him.[53] Such a utopian view of discipline, subordination and training did not, of course, always square with experience (apprentices talked back, disobeyed, even ran away), but that does not reduce the importance of the institution in the prescriptive edifice of mastership.

The institution of apprenticeship served another function, as well, for it contributed to the categorization of skill by gender. Elizabeth Musgrave's work on women in the building trades of Brittany shows that since the sixteenth century craftsmen had limited female access to training in the crafts by excluding them from apprenticeship.[54] Statutes never mention female apprentices, and no apprenticeship contracts for women have been found for these trades, and so 'skill' was defined as a property exclusively available to men. Artisans working on construction sites – roofers, masons, carpenters – monopolized the positions that required training, and relegated what women there were on these sites to the rank of untrained, and thus 'unskilled' labourers.

Faux ouvriers, *chambrelans*, 'chamberers', or *Böhnhasen*, as clandestine workers were variously called in France, England and Germany, were workers behaving as masters without guild sanction – taking orders from customers, hiring workers, and so forth – and were construed as insubordinate, indisciplined and, significantly, 'unskilled'.[55] Judging just from their technical aptitude, many of these workers would seem skilled to us, but judging from the avalanche of complaints by master guildsmen of the widespread existence of these clandestine workers and the legislation designed to regulate them, they were disparaged as 'unskilled' and, as the French language of the time tellingly put it, *sans état*, or literally 'without rank'.

Apprenticeship and the regulation of clandestine workers are clearly in part about technical aptitude, division of labour and labour supply, and they do connect with the world of production, but their significance is broader than economics, for they signal the defensive posture of exclusion taken by guild masters which was secured in part by the definition of skill. Skill, in other words, was as much a cultural construct articulating boundaries of a community defined by status and a sense of difference as an indicator of the economic capacity of a worker. For masters, skill, or the lack thereof, helped articulate a system that created a hierarchical slot (as well as a division of labour) for women, for apprentices, for journeymen, for 'roomworkers', and, of course, for themselves. Hegel perceptively pointed out that the evidence of a master craftsman's 'skill' was not his technical wizardry, but rather his membership in his guild community.[56] Such membership conferred a collective status upon the guildsman via his mastership, and granted him a sense of possessing a 'property in skill' which marked him off, in his mind, from others supposedly without it. Through mastership, the property in skill was, in theory, legally secured. Skill so construed was embedded in a male master artisan's identity which was rooted in the soil of rank and hierarchy.

If master artisans were keen on distinguishing themselves from their inferiors by a cultural definition of skill, they certainly were not the only craftsmen to do so. Journeymen, too, turned skill into symbolic capital and deployed it as part of their honour. In Germany, for example, journeymen determined that working alongside women meant a reduction in one's 'property in skill', for women's work was construed as less skilled than men's, even though, as Merry Wiesner points out, 'the actual dexterity and facility required might be more than those in men's work'.[57]

Gender, then, was a more important determinant of the distinction between skilled and unskilled activity than the technical aptitude required by the task. Gender also determined the value-laden difference between guild work, everywhere in early modern Europe associated with 'skilled labour' if it was a question of male guilds, and domestic work everywhere associated with 'unskilled labour'. The institutional world of work, in contrast to the household, was overwhelmingly male. In general, all across Europe women saw their formal, independent participation in guilds narrow from the middle ages to the eighteenth century. Although their presence was not entirely eroded, guildsmen and magistrates joined increasingly to exclude women from a range of guilds, leaving them in guilds on which society placed little social value and deemed 'unskilled'. This gender division of labour was firmly supported by patriarchal theory and takes its place in the accelerating hierarchical disciplining of society – in this case, of women – and increasingly relegated independent women to poorly paid, insecure, and politically powerless – in a word, inferior – occupations.[58]

Women were concentrated in circumscribed occupations like sewing or linen-weaving, crafts that were redolent of the household, ascribed low status, and tagged by men with the badge of 'unskilled'. This illustrates well how 'skilled' activity had become a male preserve. Jean Quataert offers a German example. Under the pressure of commercial capitalism, she argues, guildsmen in Central Europe by the mid-seventeenth century saw their markets shrinking and, in response, 'limited women's productive roles in goods manufacture'.[59] Before the Thirty Years' War, she asserts, guildsmen had disparaged household production because it was non-guild, but after the war guild prejudice against household work was laid across a gender grid. Now, after 1650, commercial capitalism drove guildsmen to adopt a gender order that reserved to the men control over production for exchange by excluding women from guild status and relegating them to the sphere of unskilled labour. Household production in turn became feminized, 'inextricably linked with women's work in the eyes of the threatened craftsmen'. Such work was 'disesteemed' in a world 'obsessed by status', and the status of honourable, skilled work thus became increasingly defined by men by its distance from household, and thus feminized, production.[60] Thus, in Quataert's picture, the lines of sex role definition causally intersected with the alienation of the household from the market economy, and played a determinant role in the meaning of skill.

Conclusion

It seems to me that there is no Archimedean point of cultural analysis, no point outside culture from which to analyse it 'objectively'. What is needed, then, is an analytical theory that enables an examination of culture meaningfully constituted by human agents which accounts for experience without privileging materiality. Such a theory must also allow the definition of social units of analysis without essentialism.

If cultures are meaningful orders of persons and things, then cultural history might profit by attending to individuals and groups of individuals purposefully engaged in the constitution of meaning through dialogical relationships of symbolic exchange with and within an 'always already' culture. Furthermore, the constitution of meaning is a cognitive operation inextricably associated with distinction and differentiation, where the sense of social being is fashioned – often by contestation – by exclusion from and inclusion in groups. Such a methodological approach melds the notion of culture as meaningful orders of persons and things with the notion that the constitution of meaning occurs within a system of differences constituted by inherently dynamic symbolic inscription and exchange. It thus takes seriously the 'linguistic turn' that has been so important to theorizing cultural history while avoiding the great pitfall of that turn, the dissolution of referentiality.[61]

I have suggested that by 'denaturalizing' labour we free ourselves to consider other dimensions of artisan identity which may have been as, perhaps even more, constitutive of it. Culture subsumes all human activity and is 'always already', so analysis cannot detach one activity (labour, for example) and posit it as prior to culture without tautological or reductive results. Instead, it seems to me, cultural analysis of artisans or any other social group should be 'reflexive', where the varieties of human activity are understood in dialogical relation with one another. Cultural history practised in this way, then, in this case the history of artisan culture, can bring together identity and social stratification while demonstrating their interconnectedness. It can thereby address one of the most intractable problems that has confronted cultural historians over the past ten years.

Notes

1. For their helpful comments and suggestions, I would like to thank Dean Ferguson, Margot Finn, Christopher Friedrichs, Nancy Gabin, John Lauritz Larson, Whitney Walton, Melinda Zook, the participants in the Vann Seminar at Emory University, and the participants in the Workshop in Early Modern European and British studies at Harvard University.
2. Useful introductions to cultural history are Lynn Hunt (ed.), *The New Cultural History*, Berkeley: University of California Press (1989); and Roger Chartier, *Cultural History*, trans. Lydia Cochrane, Ithaca: Cornell University Press (1988). Richard Biernacki, *The Fabrication of Labour: Germany and Britain, 1640-1914*, Berkeley: University of California Press (1995), ch. 1, offers a critique of the leading anthropological methods that historians have applied to cultural history.
3. Emile Levasseur, *Histoire des classes ouvrières et de l'industrie en France avant 1789*, 2 vols, Paris: Arthur Rousseau (1900-01); George Unwin, *Industrial Organization in the 16th and 17th centuries*, Oxford: Clarendon Press (1904); David Landes, *Prometheus Unbound*, Cambridge, MA: Harvard University Press (1969).
4. Peter Burke, 'The Language of Orders in Early Modern Europe', in M.L. Bush (ed.), *Social Orders and Social Classes in Europe since 1500: Studies in Social Classification*, London: Longmans (1992), p. 12.
5. Maurice Godelier, 'Work and Its Representations: A Research Proposal', *History Workshop*, 10, 1980, pp. 164-74.
6. Lenard Berlanstein (ed.), *Rethinking Labor History: Essays on Discourse and Class Analysis*, Urbana: University of Illinois Press (1993). Simona Cerutti, *La ville et les métiers. Naissance d'un langage corporatif (Turin, 17e-18e siècle)*, Paris: Éditions de l'École des Hautes Études en Sciences Sociales (1990); Michael Sonenscher, *Work and Wages: Natural Law, Politics, and the Eighteenth-Century French Trades*, Cambridge: Cambridge University Press (1989); Joan W. Scott, *Gender and the Politics of History*, New York: Columbia University Press (1988); and William Reddy, *The Rise of Market Culture*, Cambridge: Cambridge University Press (1987).
7. Lenard Berlanstein, 'Preface', in idem, *Rethinking Labor History*, p. vii.
8. James R. Farr, *Butchers, Bakers, and Candlestickmakers: Artisans in Europe from the Black Death to Industrialization*, Cambridge: Cambridge University Press (forthcoming).
9. Peter Sahlins, *Boundaries: The Making of France and Spain in the Pyrenees*, Berkeley: University of California Press (1989), p. 270.
10. Marshall Sahlins, *Culture and Practical Reason*, Chicago: University of Chicago Press (1976), pp. 157-8, 164; Jean Baudrillard, *The Mirror of Production*, trans. Mark Poster, St Louis: Telos Press (1975).

11. Simona Cerutti, 'Du corps au métier: La Corporation des tailleurs à Turin entre XVIIe et XVIIIe siècle', *Annales: ESC,* 43, 1988, pp. 323-52. Karl Polanyi, *The Great Transformation,* London: Victor Gollancz (1944), offers a powerful critique of ahistorical economism. On Polanyi, see Fred Block and Margaret Somers, 'Beyond the Economistic Fallacy: The Holistic Social Science of Karl Polanyi', in Theda Skocpol (ed.), *Vision and Method in Historical Sociology,* Cambridge: Cambridge University Press (1986), pp. 47-84. André Gorz, *Critique of Economic Reason,* trans. Gillian Handyside and Chris Turner, London: Verso (1989); and J. Davis, 'Rules not Laws: Outline of an Ethnographic Approach to Economics', in Bryan Roberts, Ruth Finnegan and Duncan Gallie (eds), *New Approaches to Economic Life,* Manchester: Manchester University Press (1985), pp. 502-11. For an overview of the current debate over the role of purposeful reason in human action which informs debates about economism, see Loïc Wacquant and Craig Calhoun, 'Intérêt, rationalité, et culture: A propos d'un récent débat sur la théorie de l'action', *Actes de la recherche en sciences sociales,* 78, 1989, pp. 41-60.
12. Dominick LaCapra, *History and Criticism,* Ithaca: Cornell University Press (1985).
13. Marshall Sahlins, *Culture and Practical Reason,* p. x.
14. Pierre Bourdieu, *The Practice of Logic,* Stanford: Stanford University Press (1990); Baudrillard, *Mirror of Production;* Christopher H. Johnson, 'Lifeworld, System, and Communicative Action: The Habermasian Alternative in Social Thinking', in Berlanstein, *Rethinking Labor History,* pp. 55-89.
15. Baudrillard, *Mirror of Production,* p. 33. The nature of the relationship between culture and society currently constitutes a fundamental debate among social theorists. For a useful introduction and overview, see Jeffrey C. Alexander and Steven Seidman (eds), *Culture and Society: Contemporary Debates,* Cambridge: Cambridge University Press (1990).
16. Patrick Joyce, 'The Historical Meaning of Work,' in idem (ed.), *The Historical Meanings of Work,* Cambridge: Cambridge University Press (1987), pp. 1-2.
17. Richard Whipp, 'Labour Markets and Communities: An Historical View', *Sociological Review,* 33 (14), 1985, pp. 768-91.
18. Paul S. Seaver, *Wallington's World: A Puritan Artisan in Seventeenth-Century London,* Stanford: Stanford University Press (1985); Jacques Rancière, 'The Myth of the Artisan: Critical Reflections on a Category of Social History', *International Labor and Working Class History,* 24, 1983, pp. 1-16.
19. Scott Lash, 'Pierre Bourdieu: Cultural Economy and Social Change', in Craig Calhoun, *et al.* (eds), *Bourdieu: Critical Perspectives,* Chicago: University of Chicago Press (1992), pp. 193-211. On the importance of the history of difference, Joan W. Scott,' The Evidence of Experience', *Critical Inquiry,* 17, 1991, pp. 773-97.
20. Donald Reid, *Paris Sewers and Sewermen in the Nineteenth Century,* Cambridge, MA: Harvard University Press (1991); idem, 'Reflections on Labor History and Language', in Berlanstein, *Rethinking Labor History,* pp. 39-54.
21. Marshall Sahlins, *Culture and Practical Reason,* pp. 203, 206.
22. Baudrillard, *Mirror of Production,* pp. 98-9.

23. On the historical relationship between hierarchy and difference in early modern Europe, see Robert Muchembled, *L'Invention de l'homme moderne*, Paris: Fayard (1988), esp. pp. 426-34; and Michel Foucault, *The Order of Things*, New York: Pantheon (1970).
24. For example, Carolyn Lougee, *Le Paradis des femmes: Women, Salons, and Social Stratification in Seventeenth-Century France*, Princeton: Princeton University Press (1976); Jonathan Dewald, *Aristocratic Experience and the Origins of Modern Culture, France, 1570-1715*, Berkeley: University of California Press (1993); and James R. Farr, *Authority and Sexuality in Early Modern Burgundy, 1550-1730*, New York: Oxford University Press (1995).
25. Bernard Chevalier, *Les Bonnes Villes de France du XIVe au XVIe siècle*, Paris: Aubier (1982).
26. Most historians agree that spiritual confraternities emerged first and craft guilds were grafted on to them in many cities in Europe between the late eleventh century and the thirteenth century. Steven Epstein, *Wage Labor and Guilds*, Chapel Hill: University of North Carolina Press (1991).
27. Bernard Chevalier, 'Corporations, conflits politiques et paix sociale en France aux XIVe et XVe siècles', *Revue historique*, **268** (1982), pp. 17-44.
28. Ibid., p. 36.
29. Catharina Lis and Hugo Soly, '"An Irresistible Phalanx": Journeymen Associations in Western Europe, 1300-1800', *International Review of Social History*, **39**, 1994, pp. 11-52; James R. Farr, *Hands of Honor: Artisans and Their World in Dijon, 1550-1650*, Ithaca: Cornell University Press (1988), pp. 65-75; Cynthia Truant, *The Rites of Labor: Brotherhoods of Compagnonnages in Old and New Regime France*, Ithaca: Cornell University Press (1995); Andreas Griessinger, *Das symbolische Kapital der Ehre: Streikbewegungen und kollektives Bewusstsein deutscher Handwerksgesellen im 18. Jahrhundert*, Frankfurt: Ullstein (1981); Hans-Ulrich Thamer, 'On the Use and Abuse of Handicraft: Journeyman Culture and Enlightened Public Opinion in 18th and 19th Century Germany', in Steven L. Kaplan (ed.), *Understanding Popular Culture: Europe from the Middle Ages to the Nineteenth Century*, Berlin: Mouton (1984), pp. 275-300.
30. Ulrich-Christian Pallach, 'Fonctions de la mobilité artisanale et ouvrière: Compagnons, ouvriers, et manufacturiers en France et aux Allemands (17e-19e siècles)', *Francia*, **11**, 1983, p. 368.
31. Farr, *Hands of Honor*, esp. ch. 4; Mack Walker, *German Home Towns*, Ithaca: Cornell University Press (1971).
32. Pallach, 'Fonctions de la mobilité artisanale', pp. 395, 399.
33. Merry Wiesner, '*Wandervogels* and Women: Journeymen's Concepts of Masculinity in Early Modern Germany', *Journal of Social History*, **24** (4), 1991, pp. 767-82.
34. Merry Wiesner, 'Guilds, Male Bonding, and Women's Work in Early Modern Germany', *Gender and History*, **1**, 1989, p. 128.
35. Arlette Farge, *Fragile Lives: Violence, Power and Solidarity in Eighteenth-Century Paris*, trans. Carol Shelton, London: The Polity Press (1993), pp. 268, 271.
36. Robert Darnton, 'A Printing Shop Across the Border', in *The Literary Underground in the Old Regime*, Cambridge, MA: Harvard University Press (1982), pp. 159-61, 165-6.

37. Thamer, 'Use and Abuse of Handicraft', pp. 280, 287-8; Wiesner, '*Wandervogels* and Women', p. 770.
38. Jacques-Louis Ménétra, *Journal of My Life*, trans. Arthur Goldhammer, New York: Columbia University Press (1986).
39. Bronislaw Geremek, *Le Salariat dans l'artisanat parisien aux XIIIe-XIVe siècles*, Paris: Mouton (1962), pp. 110-18; Steven L. Kaplan, 'La Lutte pour le contrôle du marché du travail à Paris au XVIIIe siècle', *Revue d'histoire moderne et contemporaine*, 36, 1989, pp. 361-412; L.D. Schwarz, *London in the Age of Industrialisation*, Cambridge: Cambridge University Press (1992), pp. 120-21, 189-91; Michael J. Neufeld, *The Skilled Metalworkers of Nuremberg: Craft and Class in the Industrial Revolution*, New Brunswick: Rutgers University Press (1989), pp. 41-5.
40. Truant, *Rites of Labor*, pp. 137, 146. Norbert Elias, *The Civilizing Process*, 2 vols, New York: Pantheon (1978).
41. Steven L. Kaplan, 'Social Classification and Representation in the Corporate World of Eighteenth-Century France: Turgot's Carnival' in Steven L. Kaplan and Cynthia Koepp (eds), *Work in France*, Ithaca: Cornell University Press (1986), pp. 176-228.
42. One can see three periods when the perception of 'crisis' was most pronounced: the late fourteenth century, the late sixteenth century and the eighteenth century when, according to Kaplan, it reached 'apocalyptic' dimensions. Steven L. Kaplan, 'Réflexions sur la police du monde du travail, 1700-1815', *Revue Historique*, **261**, 1979, pp. 17-77.
43. Geoffrey Crossick and Heinz-Gerhard Haupt (eds), *Shopkeepers and Master Artisans in 19th-Century Europe*, London: Methuen (1986); and Raphael Samuel, 'Mechanization and Hand Labour in Industrialising Britain', in Lenard Berlanstein (ed.), *The Industrial Revolution and Work in Nineteenth-Century Europe*, London: Routledge (1992), pp. 26-43.
44. Michael Sonenscher, 'Mythical Work: Workshop Production and the *Compagnonnages* of Eighteenth-Century France', in Joyce, *Historical Meanings of Work*, pp. 31-63; Geoffrey Crossick, *An Artisan Elite in Victorian Society*, London: Croom Helm (1978); Charles More, *Skill and the English Working Class*, London: Croom Helm (1980); Mary Freifeld, 'Technological Change and the "Self-Acting" Mule: A Study of Skill and the Sexual Division of Labour', *Social History*, 11, 1986, pp. 319-43; Keith McClelland, 'Time to Work, Time to Live: Some Aspects of Work and the Re-Formation of Class in Britain, 1850-1880', in Joyce, *Historical Meanings of Work*, pp. 180-209.
45. John Rule, 'The Property of Skill in the Period of Manufacture', in Joyce, *Historical Meanings of Work*, pp. 99-118.
46. Sonenscher, *Work and Wages*, p. 323.
47. Katrina Honeyman and Jordan Goodman, 'Women's Work, Gender Conflict, and Labour Markets in Europe, 1500-1900', *Economic History Review*, 44 (4), 1991, pp. 608-28.
48. See Stephen Marglin, 'What Do Bosses Do? The Origins and Functions of Hierarchy in Capitalist Production', *Review of Radical Political Economy*, 6, 1974; Harry Braverman, *Labor and Monopoly Capital*, New York: Monthly Review Press (1974).
49. The fluidity of production, the mobility of workers, the ephemeral nature of much employment, and above all the interchangeability of many technical skills among trades means that 'the names of the trades have a

very misleading precision', as Sonenscher writes; 'Mythical Work', p. 55.
50. Kaplan, 'Social Classification', p. 177.
51. A.-V. Chapuis, *Les Anciennes corporations Dijonnaises: Règlements, statuts, et ordonnances*, Dijon: Nourry (1906), p. 308.
52. Ibid., pp. 209-10.
53. Farge, *Fragile Lives*, p. 117.
54. Elizabeth C. Musgrave, 'Women in the Male World of Work: The Building Industries of Eighteenth-Century Brittany', *French History*, 7 (1), 1993, pp. 30-52. In other trades many women did serve apprenticeships, but after 1550 almost none of them were tracked for mastership. Instead, apprenticeship was an institution of cheap labour quite often, estimating from the tasks many of these 'apprentices' were expected to do, indistinguishable from the activities of domestic servants. Such service had little to do with the acquisition of a 'skill' in the way of boys. Deborah Simonton, 'Apprenticeship: Training and Gender in 18th-Century England', in Maxine Berg (ed.), *Markets and Manufacture in Early Industrial Europe*, London: Routledge (1991), pp. 227-58.
55. Farr, *Hands of Honor*, p. 61; Walker, *German Home Towns*; Steven L. Kaplan, 'Les Corporations, les *faux ouvriers*, et le faubourg Saint-Antoine au XVIIIe siècle', *Annales: ESC*, 43, 1988, pp. 453-78.
56. Antony Black, *Guilds and Civil Society in European Thought from the Twelfth Century to the Present*, London: Methuen (1984), p. 204.
57. Wiesner, '*Wandervogels* and Women', p. 777.
58. Natalie Z. Davis, 'Women in the Crafts in Sixteenth-Century Lyon', in Barbara Hanawalt (ed.), *Women and Work in Preindustrial Europe*, Bloomington: Indiana University Press (1986), pp. 167-97; Martha Howell, *Women, Production, and Patriarchy in Late Medieval Cities*, Chicago: University of Chicago Press (1986); Merry Wiesner, *Working Women in Renaissance Germany*, New Brunswick: Rutgers University Press (1986); Judith Coffin, 'Gender and the Guild Order: The Garment Trades in Eighteenth-Century Paris', *Journal of Economic History*, 54 (4), 1994, pp. 768-93.
59. Jean H. Quataert, 'The Shaping of Women's Work in Manufacturing: Guilds, Households, and the State in Central Europe, 1648-1870', *American Historical Review*, 90 (5), 1985, p. 1124.
60. Ibid., p. 1132.
61. On the 'linguistic turn' and the debate over the relationship between meaning and experience, see Scott, 'The Evidence of Experience'; John E. Toews, 'Intellectual History After the Linguistic Turn: The Autonomy of Meaning and the Irreducibility of Experience', *American Historical Review*, 92 (4), 1987, pp. 879-907; John H. Zammito, 'Are We Being Theoretical Yet? The New Historicism, The New Philosophy of History, and Practicing Historians', *Journal of Modern History*, 65 (4), 1993, pp. 783-814; and Dominick LaCapra, *Soundings in Critical Theory*, Ithaca: Cornell University Press (1989).

CHAPTER FOUR

'Broken all in pieces': artisans and the regulation of workmanship in early modern London

Michael Berlin

From the early medieval period up to the eighteenth century, a dominant feature of the economic life of the City of London was the organization of groups of craftsmen, traders and merchants into self-regulatory guilds or livery companies which in the metropolis, as in other English cities and towns, came to be a major feature of urban society. Production, distribution and supply were dominated by the livery companies, which regulated urban markets, and provided the electorate for the assemblies of urban government. This essay is devoted to examining a set of social practices which were associated with the operation of the livery companies, the powers of inspection of manufactured goods, in order to see what can be learned about the changing social and economic forces which shaped metropolitan society, and the role of artisans inside and outside livery company organizations. By looking at the exercise of these policing powers, especially at those points at which consensus over their enforcement broke down, it is hoped to gain insight into how artisans viewed the exercise of their productive capacities during a period when the process of economic, social and technological change was gradually dissolving many of the tacit assumptions upon which the system of regulated craft production was based.

At the centre of this system of supervision of production were the powers of police which the guilds and livery companies exercised over the manufacture and sale of goods. Such powers, alongside the control of access to membership via the system of apprenticeship, were part of the armoury of regulations by which members of the livery companies sought to control production and distribution in their own interests. Primary to this was the power to control the quality of the products of artisanal labour via periodic public inspection.[1] This mechanism of control was also given sanction in civic by-laws which sought to protect urban consumers from 'false and deceitful wares'. The protection of urban consumers from fraud was buttressed by the paternalist outlook of the early modern English state, which vested powers in guild and city officials to inspect standards as a means of upholding the Crown as guarantor of the urban consumer against fraud and deceitful practices. Designed to fufil the duty of the monarch to act in the best interest of the subject, the Tudor state's involvement in the regulation of urban markets also helped to enforce national policies designed to stimulate exports, and to monitor production in industries in which the Crown had an interest.

Despite much scholarly endeavour, the chronology and extent of guild control of the English urban economy during the early modern period remains unclear. While most historians would agree that the livery companies and other guilds continued to exert their powers well into the sixteenth century, they present very different accounts of the timing of the diminution of these powers. Some date the fading role of the guilds and livery companies from the early seventeenth century, while others have suggested the later seventeenth and early eighteenth century.[2] Yet a convincing case could be argued for dating the decline of guild powers of control from the mid-eighteenth century onwards.[3] If this chronology is unclear, the explanations offered for the breakdown of guild control tend towards teleological assumptions about the inherent incompatibility of guilds with economic growth, technological change and the development of capitalism. However, the history of waxing and waning of powers of inspection of the London livery companies suggests that the breakdown of guild control came through the conscious decision of those who controlled the companies, rather than because of a failure to adapt to an unseen historical dynamic.

The history of the London companies stands out from the experience of other English towns. Throughout the early modern period, in contrast to many other urban centres, London's economy enjoyed continued expansion. The population of the metropolis, which grew from somewhere in the region of 50 000 inhabitants at the beginning of the sixteenth century to well over one million by the mid-eighteenth century, was the fulcrum of the city's growth.[4] As London's population expanded, the built-up area of the metropolis more than doubled in size as the open fields surrounding the medieval city gave way to speculative building on a grand scale. Population growth reflected the geographical concentration of wealth in a metropolitan-based landed elite, which spent increasing amounts of time and money in the metropolis in pursuit of conspicuous consumption. This consumption was catered for by London's artisans who, over the course of the period, developed an enhanced reputation for the quality of their skills. The high wages of London's artisans attracted skilled labour from all over the British Isles and beyond. At the same time, the process of enclosure and other changes in the agrarian economy, which affected the employment opportunities of swathes of the rural population, provided the metropolis with a steady source of migrants. In the course of the early modern period, as older corporate towns struggled to find a place alongside emerging centres of specialized proto-industrial development, London became the great centre of artisanal production, the single most concentrated place for the making and marketing of goods in Britain, and by the mid-eighteenth century possibly the largest manufacturing centre in Western Europe.[5]

This growth presented opportunities as well as dilemmas for London's livery companies. The livery companies were an integral part of the economic and social regulation of the urban environment from the early medieval period and, so long as the size of the city's population remained consonant with the powers of the livery companies, they continued to do so. But as the population expanded in the later sixteenth century there were increasing signs of strain. The growth of the built-up area increased the potential scope of responsibilities of the livery companies, while weakening their exclusive claims to regulate urban markets for goods and services. The expansion of the built-up area outside the control of the guilds made enforcement of guild regulation increasingly difficult.

At the same time the continuing expansion of artisan-based industries changed the social composition of the livery companies. The increasing use of sub-contracting to outworkers gave rise to a class of employers who controlled supplies of raw materials and, in the case of textiles, equipment such as looms, and a much larger class of artisans, small-scale master craftsmen and journeymen who depended on sub-contractors for work and materials. The traditional ideal of livery company membership, that of the self-sustaining master artisan, was gradually being undermined by the development of capitalized manufactures. For most of the sixteenth century the social and economic divisions within the livery companies were contained by a set of institutional sanctions and paternalistic attitudes which reduced the likelihood of social conflict.[6] The interests of artisans were represented in tiers of membership within the companies known as yeomanry. Demarcation disputes between livery companies worked against the formation of social identities which transcended horizontal divisions.

The early seventeenth century was marked by the proliferation of new incorporations of artisans, often breaking away from older organizations. Some 27 new incorporations were formed between 1600 and 1640. The creation of these new incorporations was actively pursued by the Crown, which encouraged courtier monopolists to act as patrons for groups of artisans. Though appearing to act in the interest of London's craftsmen, the Crown's policy of encouraging new incorporations was designed to stimulate revenue and reward favourites.

The period was marked by a series of test cases between 1599 and 1614, culminating in the famous judgement in Tolley's case, which undermined the legal controls which livery companies claimed to exert.[7] The companies responded with campaigns to extend geographical areas under their nominal control, by extending powers of inspection to the metropolitan suburbs. They also attempted to persuade all those artisans practising a trade to enrol in their respective companies.

Though the expansion of London's trade and population in the early seventeenth century undermined the position of the companies, they continued to act as the main focus for artisan organization. During the political upheavals of the 1640s and 1650s the artisan element within the companies agitated for greater enforcement of trade controls, in the form of inspections, apprenticeship regulations, and the enrolment of non-freemen. This agitation has been seen as crucial to the inculcation of democratic ideas through collective self-regulation.[8] Though the artisan movements within the livery companies were unsuccessful, London's artisans continued to look to the livery companies as an effective means of exerting collective control. Developments in the later seventeenth century were to further undermine the position of the livery companies, and of the artisans within them. The Great Fire of 1666 led to the temporary lifting of the requirements of membership in the livery companies, a measure intended to stimulate the quick rebuilding of the charred city. But this amnesty from guild membership led to an influx of non-guild labour, a development which seriously weakened guild control. Though the livery companies attempted to reaffirm their powers in the 30 years after the fire through new charters of incorporation and the extension of local by-laws, the further expansion of the built-up area of the metropolis in the early eighteenth century meant that the area to which powers of inspection might apply was more and more densely populated.

It was not only the demographic and spatial expansion of the metropolis which sapped the powers of the guilds. The development of divergent social forces in the metropolitan economy, within and between the various livery companies and occupational groupings, meant that the enforcement of these powers became a frequent source of conflict. Competing groups of craft producers used these powers in claims to dominate areas of production and to enforce the occupational homogeneity of each livery company. Within many of the larger craft-based companies, such as those involved in textiles, metalworking and leatherworking, the right to supervise the results of artisanal production became central to the increasingly unequal relationship between an occupational hierarchy of merchants and wealthier master craftsmen who dominated the hierarchy of office-holders and artisans who made up the rank and file membership. In the larger craft-based industries the social distinctions between artisan producers and mercantile elites had been institutionalized in the internal division of livery company government into a hierarchy of officers and subordinate bodies. Craftsmen tended to become represented by an inferior body known as the yeomanry. This division grew sharper as the seventeenth century progressed, and suburban growth and the proliferation of outwork diminished the economic independence of the artisan. To this increasingly disenfranchised class, standards of workmanship were a measure of social standing. The livery company's claim to regulate standards was thus a means of collectively protecting the exercise of skilled labour.

The right to enforce prescribed standards of workmanship was thus a fundamental feature of livery company government, stipulated in the earliest medieval ordinances and charters of incorporation.[9] Enforcement took the form of so-called 'searches': public perambulations of the city streets by guild officials who visited workshops, warehouses and places of sale, inspecting the quality of goods and raw materials. Regulations governing the right to search empowered officials to exact a variety of sanctions against offenders; these included the right to seize faulty goods, to levy fines, to inflict punishments of imprisonment and to destroy the seized products. Searches were conducted by specially appointed officials, often representatives or wardens of the subordinate bodies within the livery companies, the yeomanry. In some companies specialist 'viewers', 'advocates', 'witnesses' and 'searchers' were also added, men whose acknowledged expertise in the handicrafts made them suitable as arbiters of their colleagues' products.

Livery company ordinances stipulated that searches be held at regular intervals throughout the year, timed to coincide with the various meetings at their respective halls. The frequency of inspection varied from company to company, some conducting searches before the four quarterly meetings of the whole company while others held the search up to 12 times a year. Many companies held their searches at the beginning of the annual celebration of Bartholomew Fair in late August. On these occasions livery company officials, accompanied by legal representatives of the city corporation, dressed in official costume and carrying insignia, processed from their halls through the city and the site of the fair in Smithfield, stopping at workshops and stalls. The timing of the perambulation, coinciding with the official opening of fairs and market days when buying and selling were at their height, helped to emphasize the public assertion of the maintenance of shared standards of workmanship.

The public affirmation of the right to search was also implied by the routes taken by the company officials. Responsibility for the search was apportioned according to specific areas of the city's jurisdiction, which reflected the major concentrations of particular groups of artisans and traders. For purposes of the search the city was divided in geographical zones or 'walks', and during the inspection attention would be paid to quality control and the qualifications of the craftsmen in the workplace. The Clockmakers' search consisted of four 'walks': 'the City', 'North-west', 'South-west' and 'Eastward', which covered the major concentrations of working clockmakers throughout the built-up area. The Coopers' Default Books show that a party of eight members of the company, clearly relying on information provided by informers, would set off to make a 'strict search into the most likely places where wee were advised abuses may have been committed'. The routes of their search were varied every three months, although on each occasion they managed to visit about thirty different workplaces. As the metropolitan area expanded, the companies sought with differing measures of success to extend the area covered by the search. In those companies where the powers of search were associated with exercise of important legal functions devolved on to the livery company by the Crown, the area of search was extended still further to cover provincial manufacturing centres. The Goldsmiths' and the Stationers' Companies respectively were granted powers by the Crown to inspect their trades thoroughout England in order to test (or assay) the fineness of gold and silver and to help the state monitor the distribution of printed books.[10]

The search dealt with a wide variety of offences not exclusive to 'quality control': working or trading by 'aliens' and 'foreigners' who did not belong to the livery companies (who hence had no right to trade openly in the city limits), the employment of numbers of apprentices beyond the limits prescribed in company ordinances, and the monitoring of the use of new techniques and inventions which threatened the reputation of existing skilled artisans. In the hands of the wardens of yeomanry the search became a mechanism for exerting collective control over production. As such it became a source of contention and negotiation in those trades where a widening social distance existed between merchants and artisans. In the case of the Clothworkers' Company, a social equilibrium was uneasily maintained between the interests of artisans and merchants in which the search was regularly enforced as a means of maintaining the social stability of the company.

A central element of the enforcement of the powers of search was the ritualized public destruction of faulty products. This aspect of the search took a variety of forms. Confiscated goods were subject to quasi-judicial proceedings in which prosecution and punishment were as much directed at the inanimate object as at the person. Destruction took the form of judicial punishment meted out in markets and thoroughfares, outside shopfronts and before assemblies at the livery company halls. The turners' ordinances decreed that 'faulty commodities ... sold to the great slander of the Misterie' were to be viewed by the master, wardens and two assistants, though if any chair were bought before it was searched, or 'if any of the seat lifts of the said chairs are made of any wood than Ash, they may deface such chairs as heretofore they have usually done'. What this entailed may be gleaned from an additional by-law of 1630, which stated that thereafter it was necessary to try the quality of the chair by means of two feet instead of one.[11] Basketmakers paid some £7 15s 6d to several 'witnesses' who carried defective wares to Bartholomew Fair where, after being duly condemned by a 'jury', the offending goods were 'burnt and consumed'. The Spectaclemakers' officers held similar 'trials' at the Mayor's Court at Guildhall. As a description of their proceedings against 22 spectacles seized from the shop of Elizabeth Bagnall in 1671 makes clear, the form of the search closely followed criminal procedure. After a jury at the Mayor's Court found them to be 'badd and deceiptful', the glasses were 'by the judgement of the Court condemned to be broken defaced and spoyled both glasse and frames the w[hi]ch judgement was executed accordingly in Canning Street on the remayning parte of London Stone where the same were with a hammer broken all in pieces'.[12]

A standard justification given in livery company records for the rights of search interwove the interests of producers, in the form of the need to uphold collectively possessed public reputation for quality of manufactured articles, with the interests of consumers who had the right to be protected from faulty or fraudulently made goods. Company ordinances emphasized this combination of interests, promulgating the inspection of goods in terms which combined concern with the debasing of artisanal skills with the protection of the public from fraud. The 1571 petition of 14 companies representing the 'handicrafts' to the Common Council of London for action to see the powers of search vigilantly enforced used terms which emphasized this social compact between producers and consumers:

> In olden times past, when the companies of artificers and handicraftsmen of this city reserved to themselves the only use, trade, and exercise of their several arts and handicrafts, the things then pertaining to the said arts were truly workmanly and substantially made and the Queen's Majesty's subjects well and truly served thereof.[13]

The sanction of destroying finished products rested on the desire to uphold the collective reputation of the trade threatened by faulty materials or poor workmanship. The by-laws of the Weavers' Company empowered the search to look out for cloth which was 'unartificially, insuficientlie or falslie wrought'. The Clockmakers' Company justified the defacing of a gold watch-case found by the officers of the search to be made of 'coarse and unwarrantable gold' because 'great fraud would be put upon ye person who should happen to buy it, and an abuse and disparagement redound to the Art, and all good and honest Artists.'[14] Such judgements placed great emphasis on upholding the public reputation of the collective working standards of the companies. Working at the top end of the luxury trade, clockmakers were particularly concerned about the public reputation of their products. Apart from preventing both unqualified aliens and strangers, as well as those who had not served their seven-year apprenticeship, from plying their trade, their search committee

> had power to search and view all productions of the Art made within the Realm or imported for sale, and to seize and break work unlawfully made or composed of bad or defective materials, or in any way faulty; to carry the same to the Company's Hall or meeting place to be adjudged, and if condemned, to be broken [and] to break open any place, if refused admission, to search it, and to seize work and tools therein concealed ...

The records, dating from the company's foundation in the early 1630s, show that a number of goods were declared to be 'deceitfull' and 'unserviceable', to be then 'defaced and broken that [they] might not be put to sale to deceive the people'.[15]

Considerable flexibility was employed in the handling of offenders. Livery companies showed discretion in sentencing, imposing smaller fines in appropriate cases. As Ian Archer has pointed out, the system of treating those who willingly submitted to the judgement of the search with leniency, while dealing more harshly with the recalcitrant, helped to secure acceptance of the practice by the individuals involved.[16] Different reasons were given for their opposition. The wire-drawer Philip Washbourne argued that he would lose his trade secrets to other craftsmen and 'specifically asked that those who used engines should not be allowed into his work-room', a request to which the searchers acquiesced.[17] Some chose more drastic means of obstruction, such as when a Mr Fennings of the Goldsmiths' Company 'threatened to shoot one of the wardens through the head'.[18] However, even quite violent resistance to the officers of the search sometimes met with comparative magnanimity if the offender ultimately submitted to the authority of the search. In such instances, a display of obeisance served to bring the offender back within the sanctity of the brotherhood. In 1623 the Basketmakers formally forgave and remitted half the fine imposed on one man who had violently assaulted and 'fetched blood' of one of their officers in the exercise of the search 'forasmuche as hee confessed and acknowledged his said fault with an expression of hartie sorrowe, and humblie upon his knee submitted himselfe'.[19]

In the case of the Clockmakers and Spectaclemakers, submission involved the active and voluntary participation of offenders in the defacing and destruction of their own products. The Clockmakers' Company provides an interesting example of the way in which the medieval guild organization was perpetuated in the early modern period, incorporating the new trades and inventions of the seventeenth century. The company was successful in incorporating practitioners of skills such as the makers of mathematical instruments. The enrolment of 19 mathematical-instrument makers in 1667 was followed by the enforcement of the right of search of the makers of these instruments in the city. The seizure of rules and measures from several shops followed. The offenders had their products compared to official standard measures kept at Guildhall, after which they submitted to the company's rule by breaking one of the instruments themselves before having the remaining seized stock returned and solemnly promising amendment before the goods were to be exposed for sale.[20]

Enforcement was geared to the maintenance of consensus within the companies. To this end, in the larger companies, with mixed artisan/trader/merchant memberships, the governors of the companies representing the merchant elite enforced the search as a means of retaining the subservience of artisans to the hierarchy. The public nature of the search, with its ritual of humiliation and submission, followed by reintegration into the group, gave form to a system by which groups of artisans subordinated their interests to a distinct social hierarchy in return for the faithful performance of livery company elites of their duties as governors. So long as the livery company governors periodically enforced company statutes, especially those which gave artisans a stake in the control of the traditional means of upholding the reputation of their trades, then the hierarchical system of guild government could count on wider support.

This system of enforcement, coercion underscored by ritualized communal sanctions, persisted so long as the delicate balance of social forces and shared attitudes was maintained. Most of the London livery companies continued to carry out searches in the later seventeenth century. But there were signs as the period progressed of increasing unwillingness and inability of the men who controlled the machinery of the search to carry out its functions. Changes in patterns of work organization, often but not exclusively connected with the introduction of labour-saving technology, posed a challenge to the moral imperatives implicit in the right of search.

These changes were met with agitation and violent protests by artisans, both within and without the livery company environment. Historians have long recognized that early modern forms of popular protest have to be understood not as the spasmodic reactions of unthinking 'mobs' to the prospect of economic change, but as forms of protest subtly geared to achieving specific ends, or in Eric Hobsbawm's memorable phrase, 'collective bargaining by riot'.[21] Early modern popular protest drew on a rich symbolic language derived from urban and rural seasonal customs, rituals of humiliation and parodies of urban ceremonies. The complex forms of protest used by the early modern crowd often appropriated the rituals of urban government, using the symbolic forms of justice and punishment as a means of resisting the adverse affects of economic and social change.

One such instance of ritualized protest is provided by developments in the metropolitan textile industry. The increasing use in London, from the early seventeenth century onwards, of so-called Dutch or engine looms which produced ribbons, tapes and hems, became the focus for agitation by artisans within the Weavers' Company. The new type of loom was capable of producing up to eight times the amount of cloth produced by the traditional hand loom and thus posed a direct threat to the livelihoods of the working weavers who made up the rank and file of the Weavers' Company. In the agitation against the engine looms, the assertion of the powers of search became a resource for the assertion of the rights of artisans to exert collective control over the running of the guilds. This movement culminated in the campaign for direct elections to the governing body of the company during the revolutionary upheavals of the late 1640s and early 1650s.

During this period, complaints about the spread of the engine looms by the inferior yeomanry brought demands that the governing body of the Weavers use the powers of search to check the new devices. Yet the divergent social and economic interests of the governing body of the company, composed of silkmen, wholesalers and master weavers on the one hand, and the rank and file of the yeomanry, composed of small masters and journeymen weavers on the other, meant that the powers of search were only haltingly used to check the spread of the new devices. Requests from the Weavers' yeomanry to control the use of engine looms met with increasingly indifferent responses from the company governors and the Corporation of London. During the 1620s and 1630s, the Weavers' yeomanry combined demands for the vigilant enforcement of the search with appeals to the company governors, the Corporation, Parliament, and the Crown for action to limit engine looms. In keeping with a policy of intervening in the organization of industry in favour of groups of artisans, the regime of Charles I directed that the innovation be restricted. The spread of engine looms may have been halted by the disruption caused by the Civil War, but during the Restoration the demand for clothing, especially the colourful and cheap tapes and ribbons produced by the weavers, led to a renewal of complaints from the yeomanry about the increasing use of the looms. The yeomanry formally requested that the company present a petition to Parliament for a bill to restrain the looms. Yet these appeals received only lukewarm support from the governors of the company, though a petition was presented to the House of Commons. But in 1671 the master manufacturers who used the engine looms successfully lobbied for the bill to be dropped.[22]

This agitation against the engine looms culminated in the violent scenes of August 1675, when several hundred weavers in the textile-making districts of the capital marched on the workshops using the engine looms.[23] The offending devices were taken from the workshops and burnt at the doorways of workshops and in the public highways. The location for this destruction suggests that the participants were acting according to a certain set of perceived norms of behaviour derived in part from the corporate regulations designed to prevent fraud and poor workmanship which lay behind the powers of search. Earlier complaints about the engine looms by the yeomanry had echoed the justification of the powers of search. Complaints about the tapes and ribbons produced by the new looms spoke of these products as 'ill wrought and deceytfully made, to the great prejudice of the buyer, and hurt of the commonwealth'. Violence was restricted to destruction of the looms and the cloth being made: one of the rioters told how a message had spread to 'meddle only with the looms' while another description noted how the disorders had taken place 'without any other violence than barely to fetch out the engines and burn them by the doors'.[24] One master manufacturer, Thomas Bibby, who lost four looms to the crowds, had previously abused the officers of the search when they were investigating engine looms in the late 1660s.[25] The weavers involved appeared to have a purposeful sense of the impact of the new technology: one of the weavers examined after the riot had calculated precisely the potential loss of earnings due to the introduction of the new looms.[26]

Other aspects of the weavers' insurrection appear to have drawn on the sense of shared assumptions about the exercise of skills. The organization of the rioters in territorial 'divisions' (designated by the different-coloured feathers and ribbons which they wore in their hats), and the degree of sympathy for their actions shown by local militia officers, point to a perceived legitimacy of the machine-breaking as a means of upholding the traditional rights of search. The enforcement of the search helped provide the symbolic language in which artisanal skills were defended as a collectively maintained property right which all those engaged in production had an obligation to maintain.[27] The presence of a number of women amongst those arrested during the riots demonstrates the continuing importance of their role in household-centred craft production and that this defence of artisanal skills had not at this stage become an exclusively male preserve.[28] The machine-breaking was thus not an irrational response to technological change so much as a calculated attempt to maintain living standards in the face of a direct challenge. The rioters were acting where the Weavers' Company, the Corporation and Parliament had failed to act.

The riots spurred the governors of the Weavers' Company to support a new appeal by the yeomanry of the weavers to Parliament and the Privy Council for action to limit use of the new looms. As with earlier attempts at limitation, the latest appeal foundered in the face of influential opposition from the new manufacturers, some of whom were members of the governing body of the company. The master manufacturers who came to dominate the Weavers in the late seventeenth century were more concerned with the threat posed by imported cloth. The Privy Council backed the users of the engine looms in terms which underscored the extent to which the older paternalist intervention to limit the impact of technological innovation no longer prevailed. The Council noted the adverse effect that bowing to the weavers' demands would have on the development of domestic industry:

> ... that by the same reason the Single Loom Weavers complain of the Engine Looms many [other] ... envious people will complain of Engines for Water Mills, Saw Mills, and Engines for splitting of iron, and ploughs and printing presses, and Cranes for Wharfs, and many other Ingenious, useful and profitable inventions now in England, but we doubt not but Ingenuity will find encouragement in England.[29]

While the riots did not signal a wholesale abandonment of the rights of search by the Weavers' Company, in the long term the company, along with the other livery companies, abandoned the use of these powers. The chronology of this process was uneven. Indeed, in the late seventeenth and early eighteenth century there are indications that the Weavers' governors, along with many of the livery companies, sought to use the search as one means to reassert claims to control production. The resumption of searches by the Weavers' governors may have been designed to assuage pressure from below. Yet the spread of the engine looms continued unchecked as the search was directed at apprentice labour and the regulation of un-free labour.

This revival of the powers of inspection in the late seventeenth and early eighteenth century was part of a general tendency to attempt to enrol non-freemen into the livery companies and to bring the expanding metropolitan suburbs into the jurisdiction of livery company control in the aftermath of the Great Fire.[30] To this end searches were periodically enforced well into the eighteenth century, but there are indications in the middle decades of the century of a change in attitudes. The growth of London, both in terms of the metropolitan area and the population, together with the development of complex forms of sub-tenancies, made the task of enforcing the search seem daunting.[31] For the first three decades of the eighteenth century the companies reasserted the powers of search. The Weavers continued the search on a regular yearly basis down to the mid-1720s before abandoning it entirely after 1736. This was followed by a drastic decline of new members in the company.[32]

The relinquishing of the right of search was part of a general change in the mentality of ruling elites, as expressed in the decisions of livery company governors, the law courts and Parliament, which resulted in the eventual abandonment of the system of guild control. By the middle of the eighteenth century the livery companies effectively operated as employers' organizations which, with the support of the Corporation of London, sought to negotiate with journeymen organizations over wages, hours and working conditions. For the employers within the livery companies, the exclusive privileges associated with livery company control were seen to be a burden, abused by the journeymen.[33] At the same time companies increasingly faced the possibility of prosecution for trespass in the exercise of the search as legal bodies decided that local by-laws ran counter to contemporary legal opinion concerning the inviolability of private property in common law.[34]

The mid-eighteenth century saw a series of decisions by the governing bodies of the livery companies to abrogate the powers of search. The Clothworkers' Company abrogated the right of search in the mid-1750s when its governing court ruled 'the original good designs of searches almost become obsolete and totally disregarded, and the powers of search disputed by and among the clothworkers themselves'. The Clockmakers justified giving up the powers of search in 1735 because the practice was then thought to be 'interfering with the liberty of the trade'. A similar adoption of a new language of individual 'liberty' can be discerned in the decision of the Framework Knitters' Company to abandon the search in 1753, when a committee of the House of Commons ruled that powers of inspection were considered to be 'injurious and vexatious to manufactures, discouraging to industry and trade, and contrary to the liberty of the subject'. This decision was crucial in leading to the abandonment of the search by many of the companies from the mid-eighteenth century onwards.³⁵ Though the livery companies were successful in enforcing the enrolment of new members between the 1760s and 1790s, by means of various acts passed by the Common Council of London, the livery companies' claim to exercise regulation of production became increasing hypothetical. A very few companies, such as the Saddlers', continued inspecting goods well into the early nineteenth century.³⁶

An older language of defence of traditional collective rights thus gave way to a discourse of utility based on the rights of the individual. Nevertheless the older tradition metamorphosed into a new tradition of artisanal resistance. As labour historians of the late eighteenth and early nineteenth century have shown, this older rhetoric of defence of skilled artisanal labour, derived from the ancient powers of the guilds, found resonance in the earliest combinations, journeymen's organizations and box clubs, out of which emerged the first trade unions. The language of these artisanal movements drew on the symbolism of craft solidarities as a defence of what they considered to be their traditional rights.³⁷

The moral imperative of collective control of 'the trade' by artisans in the London livery companies through the mechanism of the search perhaps helps in part to explain the motivations of later generations of machine-breakers. The semi-clandestine combinations, clubs and other journeymen organizations appropriated the language of collective control which the search had meant to underscore. There are numerous instances of later destruction of machines and material by London artisans. In the 1760s sawyers, coal-heavers and silk-weavers were involved in attacks on sawmills, hoists and looms. In the Spitalfields silk-weaving district, the late 1760s was a time of near insurrection as clandestine groups of 'cutters' set about destroying cloth woven on looms owned by employers paying below the rates set in a book of prices drawn up by the working weavers.³⁸ After several years of unrest the silk-weavers' book of prices became enshrined in law.

Later attacks on machines appear to have derived a sense of legitimacy by reference to the 'old customs' of the London artisan trades. This tradition may have influenced the campaigns of the Nottinghamshire framework-knitters for the redress of grievances. The loss of powers of the London chartered company became part of the collective explanation for the adversity of the Nottingham stockingers. In 1804 the Nottinghamshire framework-knitters attempted to use the London Framework Knitters' Company as a vehicle for suits against master hosiers who made use of non-apprenticed journeymen as cheap labour.³⁹ The nineteenth-century trade unionist and chronicler of the struggles of the framework-knitters, Gravener Henson, cited an instance of machine-breaking in London in 1710 as part of his account of the history of the trade.⁴⁰ The Luddite action against looms making inferior 'cut up' hose echoed early defences of skilled workmanship. A Nottinghamshire Luddite declaration of 1811 cited the search clause in the charter of incorporation, granted by Charles II, of the London Framework Knitters' Company as granting machine-breakers powers to destroy frames and engines.⁴¹ A sense of the language of the moral imperatives of the search can be heard in *General Ludd's Triumph*, a Luddite song of the period, which declared that the 'engines of mischief were sentenced to die, by unanimous vote of the trade'. Nottinghamshire Luddism thus may have drawn on a language bequeathed by earlier struggles of London artisans to exert collective control through the livery companies.

Notes

1. The standard account remains G. Unwin, *The Gilds and Companies of London*, London: Methuen (1908). For new interpretations of the operation of the livery companies in the sixteenth century, see Ian W. Archer, *The Pursuit of Stability: Social Relations in Elizabethan London*, Cambridge: Cambridge University Press (1991), and Steve Rappaport, *Worlds Within Worlds: Structures of Life in Sixteenth Century London*, Cambridge: Cambridge University Press (1989).
2. Unwin, *Gilds and Companies*, pp. 348-50.
3. J.R. Kellett, 'The Breakdown of Guild and Corporation Control over the Handicraft and Retail Trade in London', *Economic History Review*, 2nd series, 10, 1957-58, p. 386; M. Walker, 'The Extent of Guild Control of Trades in England, c. 1660-1820', Ph.D. thesis, University of Cambridge, 1986, pp. 81-2, 91, 112-14, 182-4.
4. For the growth of London and its economic implications during the early modern period, see F.J. Fisher, 'London as an engine of economic growth' in J.S. Bromley and E.H. Kossman (eds), *Britain and the Netherlands*, London: Macmillan (1971), pp. 3-16.
5. Peter Earle, *The Making of the English Middle Class: Business, Society and Family Life in London 1660-1730*, London: Methuen (1989), p. 18.
6. Archer, *Pursuit of Stability*, pp. 140-48.
7. Kellett, 'Breakdown of Guild and Corporation Control', pp. 383-4.
8. Norah Carlin, 'Liberty and Fraternities in the English Revolution', *International Review of Social History*, 39, 1994, pp. 223-54.
9. The exercise of these powers was extensive in the livery companies in the sixteenth century. See Archer, *Pursuit of Stability*, p. 124ff; and Rappaport, *Worlds Within Worlds*, pp. 111-17.
10. Kellett, 'Breakdown of Guild and Corporation Control', p. 386; Walker, 'Extent of Guild Control', pp. 81-2, 91, 112-14, 182-4.
11. A.C. Stanley-Stone, *The Worshipful Company of Turners of London: Its Origin and History*, London: Lindley-Jones (1925), pp. 278-9 and 123.
12. F.M. Law, *The Worshipful Company of the Spectacle Makers: a History*, London: Worshipful Company of Spectacle Makers (1979), p. 378.

13. Petition of 14 'handicraft' companies to the Corporation, 1571 in Guildhall Library, Merchant Taylors' Company Court Minutes, I, fos 257-8, cited in Rappaport, *Worlds Within Worlds*, p. 111.
14. S.M. Atkins and W.M. Overall, *Some Account of the Clockmakers of the City of London*, London: privately printed (1881), p. 240.
15. Ibid., pp. 231 and 236.
16. Archer, *Pursuit of Stability*, p. 127.
17. E. Glover, *The Gold and Silver Wyre-drawers*, London: Phillimore (1979), p. 18.
18. W.S. Prideaux, *Memorials of the Goldsmiths' Company*, London: printed for private circulation (1896), vol. 2, entry for 14 October 1700.
19. H. Hodgkinson Bobart, *Records of the Basketmakers' Company*, London: Dunn, Collin and Co. (1911), pp. 115-16.
20. Atkins and Overall, *Clockmakers*, pp. 326 and 237; Guildhall Library Manuscript 2710/1, fos 185, 187, 189.
21. Eric Hobsbawm, *Labouring Men: Studies in the Development of Labour*, London: Weidenfeld and Nicolson (1964, 1968 edn), pp. 5-22. The most eloquent exposition of the culture of artisan protest during the eighteenth century is E.P. Thompson, *The Making of the English Working Class*, London: Gollancz (1963).
22. Alfred Plummer, *The London Weavers' Company 1600-1970*, London: Routledge and Kegan Paul (1972), pp. 162-9.
23. R.M. Dunn, The London Weavers' Riot of 1675', *Guildhall Studies in London History*, 1, 1975, pp. 13-23; Tim Harris, *London Crowds in the Reign of Charles I*, Cambridge: Cambridge University Press (1986), pp. 189-200.
24. Harris, *London Crowds*, p. 196; *Historical Manuscripts Commission, 7th Report*, pp. 465-6.
25. Guildhall Library Manuscript 4655/4, fo. 5 (Weavers' Court Minutes, 27 November 1666/7).
26. Harris, *London Crowds*, p. 192.
27. John Rule, 'The Property of Skill in the Period of Manufacture', in Patrick Joyce (ed.), *The Historical Meanings of Work*, Cambridge: Cambridge University Press (1986), pp. 99-118.
28. Harris, *London Crowds*, p. 193; Rule, 'Property of Skill', p. 108ff.
29. A.P. Wadsworth and J. de L. Mann, *The Cotton Trade in Industrial Lancashire, 1600-1780*, Manchester: University of Manchester (1931), pp. 102-3.
30. Kellett, 'Breakdown of Guild and Corporation Control', pp. 382-5; L.D. Schwarz, *London in the Age of Industrialisation: Entrepeneurs, Labour Force and Living Conditions, 1700-1850*, Cambridge: Cambridge University Press (1992), pp. 201-21; K.D.M. Snell, *Annals of the Labouring Poor. Social Change in Agrarian England, 1600-1900*, Cambridge: Cambridge University Press (1985).
31. Walker, 'Extent of Guild Control', p. 183.
32. Plummer, *London Weavers*, pp. 53-5.
33. C.R. Dobson, *Masters and Journeymen: a Prehistory of Industrial Relations 1717-1800*, London (1980), pp. 47-59.
34. Plummer, *London Weavers*, p. 55.
35. Kellett, 'Breakdown of Guild and Corporation Control', p. 392.
36. Ibid., pp. 393-4.

37. Rule, 'Property of Skill'; Iorweth Prothero, *Artisans and Politics in Early Nineteenth Century London: John Gast and his Times*, Folkestone: Dawson (1979); cf. Maxine Berg, *The Age of Manufactures: Industry Innovation and Work in Britain 1700-1820*, Oxford: Basil Blackwell (1986).
38. Peter Linebaugh, *The London Hanged: Crime and Civil Society in the Eighteenth Century*, London: Allen Lane (1991), pp. 256-87. For an analysis of the connections between debates in natural philosophy and machine-breaking in this period, see Alan Q. Morton, 'Concepts of Powers: natural philosophy and the uses of machines in mid-18th century London', *British Journal for the History of Science*, 28, 1995, pp. 63-78.
39. J.D. Chambers, 'The Framework Knitters' Company', *Economica*, 1929, pp. 323-4.
40. Henson's account of the riot of 1710 was included in later histories, such as W. Felkin's *A History of the Machine-Wrought Hosiery and Lace Manufactures* (1867), though the details of the action are unverified.
41. J.L. Hammond and Barbara Hammond, *The Skilled Labourer 1760-1832*, London: Longman (1919, 1995 edn), p. 259. For 'trials' and other ritualized elements in West Country machine-breaking riots, see Adrian Randall, *Before the Luddites: Custom, Community and Machinery in the English Woollen Industry, 1776-1809*, Cambridge: Cambridge University Press (1991), pp. 102-3, 194-5.

CHAPTER FIVE

The aristocratic *hôtel* and its artisans in eighteenth-century Paris: the market ruled by court society*

Natacha Coquery

In aristocratic court society, which took the king as its model, the magnificence of an aristocrat's residence stood as evidence of his rank and as a forceful statement in the competition for prestige. The lavishly decorated stately home of the nobleman - for which we shall use the contemporary French term *hôtel* - was an expression of his social distinction, good taste and aspiration for power. It was a stage, and one that was forever changing its sets. If one was to impose oneself it had to be by repeatedly devising new marks of distinction by which to shine, while simultaneously keeping imitators at bay, for imitation narrowed the gap which distinction was intended to maintain. In fact, the aristocracy was to exhaust itself in this race of constantly changing fashions. Nevertheless, the consequence was to stimulate an intense level of trade which may have been minor in comparison with the huge market for basic necessities, but still benefited the economy and helped make Paris the 'kingdom's universal shop'.[1]

How valid is it to see the aristocratic *hôtel* as a specific model of consumption, to see court society as the origin of what might be called a consumer society, even if not one in our current understanding of the term? This consumer society was induced not, as it is today, by an economic system driven by the search for profit, but by the social organization of absolutist power and its symbolic consequences. Court society encouraged extravagant consumption by constantly creating new needs. Social considerations came to prevail over economic ones. This explains the particularity of the aristocratic consumer market, which was not governed by the pure laws of supply and demand: credit and bonds brought other factors into play which created an interdependence between the partners. The socioeconomic analysis of the *hôtel* provides a good way to study both the social framework of consumption and the artisans and retailers tied into the elitist organization of society. The

---

*Translated from the French by Irene Graham and Geoffrey Crossick.

hôtel presents a coherent overall perspective through which to assess the interplay of economic, social and cultural elements.

The aim of this study is therefore to understand the mechanisms of distinction, the competitive strategies for signs of status and the way these interacted with the urban spatial structure; in short, to explore a model of expenditure in the real-life situation of court society. Therefore, rather than use the customary inventories and seals after death or acts of succession which tend to give a frozen image of the household at a given moment, we shall concentrate on bills from tradesmen and craftsmen, annual accounts of receipts and expenditure drawn up by stewards, monthly statements of outgoings by servants, lists of debts, legal documents and credit instruments. Together these provide a more dynamic picture, one which reveals the phenomenon of consumption in everyday practice. In formal economic terms, we are interested in flows rather than stocks, in acquisition rather than accumulation.

Whether concerned with material or intellectual goods, with everyday or luxury expenditure, the aristocratic consumer market called on a considerable number of 'traders and craftsmen, merchants, shopkeepers and artisans, all required to provide housing, dress, heating, food and transport'.[2] Here, we are particularly concerned with the spatial relations between the household as the place of consumption and the tradesmen who supplied it. Our aim is to ascertain the spatial impact of the economic activities which the *hôtel* stimulated, to measure the scope of that market, and to provide a case study in commercial geography: in other words, to provide a historical account of the encounters between consumers and tradesmen within urban space. Our object is therefore not to provide a general study of retailing and artisans in eighteenth-century Paris, but rather to explore the relations between a few highly privileged customers and their suppliers. The point of departure is the Parisian *hôtel*.

This study of aristocratic consumption is based on the papers of five families whose property was sequestered during the Revolution: those of the La Trémoille family, the Fitz-James, Fleury and Coigny families, and of Princess Kinsky, that is, a score of eminent persons plus their children and servants. The sample is small in comparison with the hundred or so families which made up the court, but a detailed study of specific cases can yield significant information which will contribute to our general understanding. Expenditure and lifestyle were factors common to all the households known at court. Indeed, it was the court which made uniform this language of appearances: from one *hôtel* to another the 'splendeurs courtisanes'[3] were alike. These five families were alike in the size of their income, their extravagance, their choice of suppliers and, finally, in their behaviour as consumers. In this way they may be seen as representative, especially as care has been taken to approach them from different angles. The position of each in the constellation of stars that was Versailles was different and they lived in different parts of Paris. Indeed, these families in many ways reflect the diversity of the court nobility. The La Trémoille family was of age-old noble standing, while the Fitz-James family traced itself back to the illegitimate son of an English king; the Coigny and Fleury families owed their more recent ennoblement to the king, the former coming from a background in finance and the latter from activities in the *Parlement*; the last, Princess Kinsky, was a characteristic member of the Viennese upper aristocracy. These differences cannot conceal their similarity, for we are dealing with the elite of the court, 'the hierarchical tip of the pyramid of the nobility'[4] who held the highest social rank in France, the 'plutocratic kernel' of the nobility, as Chaussinand-Nogaret has called it.[5] With the exception of the princess, widow of the chamberlain of the queen of Hungary who arrived in Paris in the early 1760s,[6] these were all dukes and peers who, whether of long standing or not, enjoyed the honours of the court.[7] The men had generally embarked on a military career at a very early age, barely 15 years old, and then achieved extremely rapid promotion, while the women became ladies in waiting. Except during the military campaigns when they were sent to fight in Europe, the New World or India, these courtiers divided their time between Paris, Versailles and their palaces in the provinces. Their fortunes were huge: their property was worth over one million *livres*, and their annual incomes ranged between an average of 60 000 *livres* for the Marquis of Fleury[8] and 300 000 *livres* for the Duke of Coigny.[9]

If the aristocratic market was characterized above all by its opulence, its organization was marked by the occupational diversity of its various suppliers, their geographical dispersal, and the nature of noble demand derived from the court society to which it was inextricably bound. The multiplicity of tradesmen and artisans, their distribution within the town, their economic strength as well as their difficulties, all reveal a clientèle that was at one and the same time powerful and captive, trapped in a world that was forever on show.

The scope of the aristocratic market

The archival evidence and household account records confirm that suppliers' bills provide a reliable indication of the total number of tradesmen involved. These total 1800, of whom over 1000 have a known address in Paris, and they covered some 200 different trades,[10] which may be grouped into six main sectors, covering 'food', 'horse', 'house', 'hygiene', 'clothing' and 'luxury goods'.[11] For each family and for each *hôtel*, maps were drawn to define the locational distribution of these suppliers within Paris.[12]

The issue becomes more complex when we see that, in spite of obvious links, the geographical distribution of the *hôtels* did not coincide with that of their suppliers. Parisian shopkeepers and artisans may have tended to follow their customers to the west, but their location did not simply match the geography of aristocratic residence. This evolved over time, in contrast to the much greater stability in the geography of tradesmen, with traditional concentrations persisting, such as the faubourg Saint-Antoine for the wood trades, rue Saint-Jacques for bookshops and rue Saint-Denis for haberdashers, all favoured by access to long-established market sites and privileged places such as the Louvre arcades, Quinze-Vingt, or Saint-Germain-des-Prés. The geographical distribution of tradesmen varied greatly with their trade: while artisans in building were dispersed, clothes merchants tended to be concentrated. It would therefore be wrong to see the *hôtel* as standing at the centre of a network of suppliers.

The area over which these artisans were spread covered the whole city *intra-muros* and even stretched beyond the boulevards: they could be found in the north as well as the south, in the west as well as the east. Although it is not the only factor to be considered, there is an obvious correlation between their number and their spatial distribution. The diameter of the area of Princess Kinsky's tradesmen was almost five kilometres, from the faubourg du Roule to the faubourg Saint-Antoine, from the sand-pit of Vaugirard to the faubourg Saint-Denis (Figure 5.1). The Fitz-James, La Trémoille and Coigny families drew from all districts, as did Princess Kinsky, with the sole exception of the Sorbonne *quartier*. All in all, no *quartier* contained fewer than ten merchants (Table 5.1). This distribution holds for all activities with the exception of the 'hygiene' sector (which contained very few people): the building, horse and clothing trades in particular were to be found almost everywhere. Notwithstanding certain areas of concentration, trades were clearly widely dispersed within the city, as is confirmed by the fact that of the 323 streets listed, almost half appear only once. Whatever the sector, streets containing only a single tradesman predominate.[13]

The breadth of this commercial network raises the question of the spatial relationship between demand and supply. The links can be explored by examining both consumers and their suppliers. The tradesmen were characterized by geographical dispersion and occupational diversity. Haberdashers were gathered in a few key places, but artisans in general were widely scattered, as clearly shown by the size of the distribution zone. They show the detailed sub-division of trades

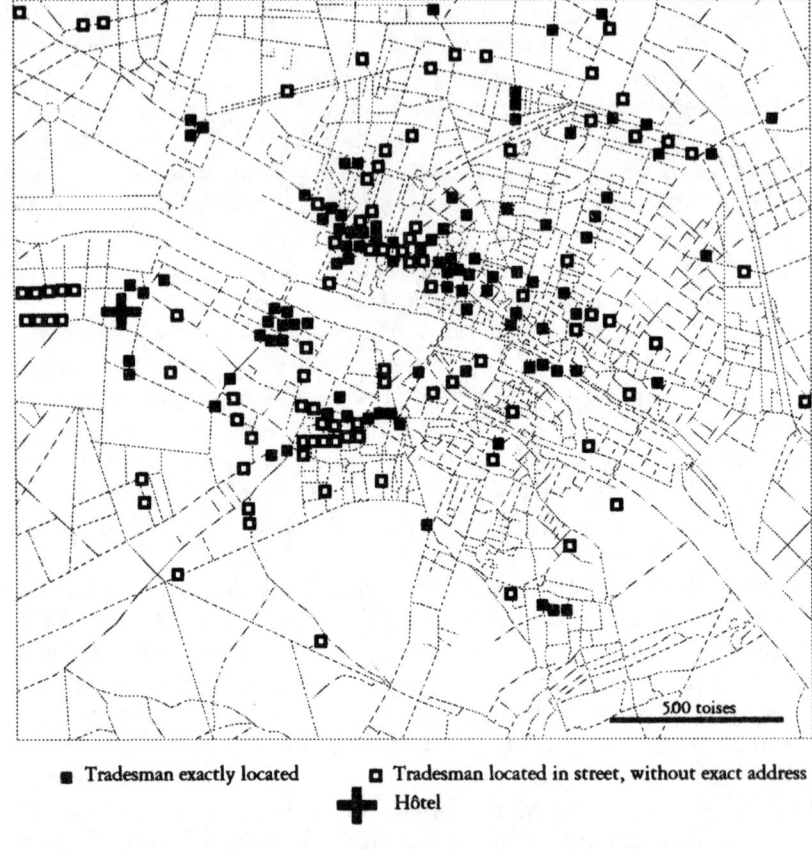

■ Tradesman exactly located   □ Tradesman located in street, without exact address
✝ Hôtel

Network of streets in Paris in 1780 following the 'plan des paroisses de Paris' drawn up by Junié, ingénieur-géographe.
Produced by Françoise Vergnesult, July 1986.

5.1 The tradesmen of Princess Kinsky, rue Saint-Dominique (1773-94), all trades

which was characteristic of the corporate system. This explains why the families used so many tradesmen: the division of labour was carried to an extreme, and codified by detailed statutes specifying minutely the tasks of each category of craftsman. The corporations, concerned to defend themselves against competition, were continuously fighting amongst themselves. They were often sub-divided into many highly specialized trade branches: carpenters were split into joiners, carpenters, coachbuilders and cabinetmakers. The haberdashers, one of the city's most flourishing guilds, boasted 20 principal branches in the eighteenth century.

The nobles, for their part, were extremely demanding, totally involved in a court society which put them under incessant pressure to find

Table 5.1 Distribution by *quartier* and trade sector of suppliers of five aristocratic families

Quartier	Food No.	Food %	Horse No.	Horse %	Luxury No.	Luxury %	Housing No.	Housing %	Hygiene No.	Hygiene %	Clothing No.	Clothing %
Hôtel de Ville					1	0.7	10	2.9			2	0.7
Place Royale	4	4.0	8	5.2	1	0.7	19	5.6			6	2.1
Marais	3	3.0	2	1.3	1	0.7	15	4.4	2	5.4	5	1.8
St-Martin	2	2.0	8	5.2	9	6.0	30	8.8			4	1.4
North-east total	9	9.0	18	11.6	12	8.0	74	21.7	2	5.4	17	6.0
St-Denis	3	3.0	10	6.5	9	6.0	22	6.5	1	2.7	8	2.8
Sts-Innocents	2	2.0	7	4.5	1	0.7	18	5.3	1	2.7	46	16.3
Halles	7	7.0	8	5.2	1	0.7	18	5.3			11	3.9
St-Eustache	7	7.0	4	2.6	7	4.7	14	4.1	4	10.8	15	5.3
Palais Royal	25	24.5	19	12.3	39	26.0	59	17.3	12	32.4	88	31.1
Louvre	8	8.0	10	6.5	11	7.3	18	5.3	2	5.4	33	11.7
North-west total	52	51.5	58	37.4	68	45.3	149	43.7	20	54.1	201	71.0
St-Germain/Prés	31	30.5	60	38.7	30	20.0	62	18.2	13	35.1	51	18.0
Luxembourg	2	2.0	13	8.4	4	2.7	18	5.3	2	5.4	8	2.8
Sorbonne					1	0.7	7	2.1			1	0.4
South-west total	33	32.5	73	47.1	35	23.3	87	25.5	15	40.5	60	21.1
Ste-Geneviève			3	1.9	6	4.0	19	5.6			2	0.7
Notre-Dame	7	7.0	2	1.3			6	1.8				
South-east total	7	7.0	5	3.2	6	4.0	25	7.3			2	0.7
Cité total			1	0.6	29	19.3	6	1.8			3	1.1
West total	85	84.0	131	84.5	103	68.7	236	69.2	35	94.6	261	92.2
East total	16	16.0	23	14.8	18	12.0	99	29.0	2	5.4	19	6.7
North total	61	60.5	76	49.0	80	53.3	223	65.4	22	59.5	218	77.0
South total	40	39.5	78	50.3	41	27.3	112	32.8	15	40.5	62	21.9
Total	101	100.0	155	100.0	150	100.0	341	100.0	37	100.0	283	100.0

Note: Suppliers with the same address, such as widows, sons, other successors in the business, have been counted only once.

craftsmen whose skill and creativity would enable them to provide a fitting display of their rank (for their distribution, see Table 5.2). They used increasing numbers of tradesmen, changed them frequently, and were willing to look far from their *hôtels* if necessary to find those whose talents would help them shine at court. Among the most numerous were some 100 cloth merchants, 60 tailors and saddlers, some 50 fodder merchants, upholsterers, carpenters or cabinetmakers, roughly 40 wine merchants and painters, and about 30 laundresses, hirers of coaches and horses, ironsmiths and glaziers. The choice of tradesmen, and hence the spatial relationship between the nobles and their suppliers, was thus dictated by what Roche has called 'the culture of appearances'.[14]

This distribution brings two other factors into relief. The first is methodological. The high-quality Parisian trades are well represented here: 1000 merchants constitute a good sample. They permit a reliable assessment of the distribution of those who worked in the capital to serve elite consumption. The second issue is analytical. The location of the dwelling, although of some importance, did not in itself determine the choice of tradesmen. The decentralized position of the *hôtel* emphasized this: Princess Kinsky lived on the farthest west side of the *noble faubourg*: no matter which sector is considered, her home was never in a central location (Figure 5.1). The same applies to the majority of other *hôtels*, all of which were situated, as was then required by fashion, on the western side of the town: the Fitz-James family in the rue de Grenelle or Saint-Florentin, the La Trémoille family in the rue Saint-

Table 5.2 Distribution by trade sector of the suppliers of five aristocratic families

Trade sector	All families		Kinsky		Fleury		Coigny		Trémoille		Fitz-James	
	No.	%	No.	%	No.	%	No.	%	No.	%	No.	%
Food	238	13.3	36	12.6	10	3.5	89	18.9	54	11.0	64	12.0
Horse	287	16.0	14	4.9	64	22.5	67	14.3	115	23.5	73	13.0
Luxury	219	12.2	33	11.6	31	10.9	58	12.3	71	14.5	57	10.0
Housing	496	27.2	107	37.5	60	21.1	123	26.2	114	23.3	144	27.0
Hygiene	85	4.8	5	1.8	14	4.9	26	5.5	18	3.7	29	5.0
Clothing	464	25.9	90	31.6	105	37.0	107	22.8	118	24.1	158	30.0
Total	1789	100.0	285	100.0	284	100.0	470	100.0	490	100.0	525	100.0

Note: Each supplier has been counted only once.

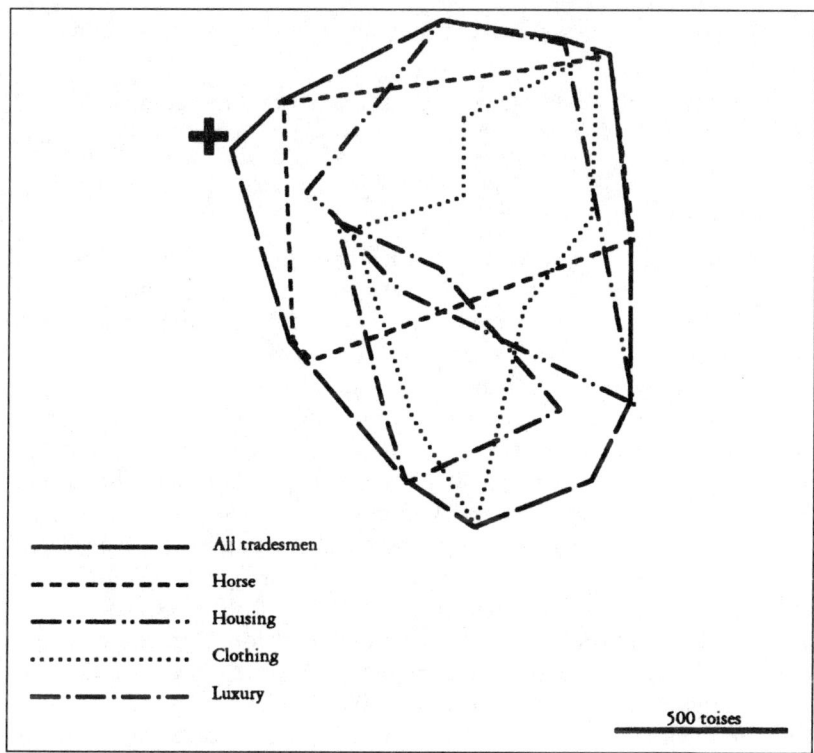

5.2 Distribution of the Coigny family's tradesmen, porte Saint-Honoré (1770–75)

Dominique, or the Fleury family in the rue de Bourbon; as for the Coigny *hôtel*, at the Porte Saint-Honoré, it was situated outside the commercial area which it actually created (Figure 5.2).

Princess Kinsky's passion for decoration and the duc de La Trémoille's love of horses provide two examples of these practices, examples which will help to explain the breadth of artisanal space within Paris. Other examples could equally have been taken, such as expenditure in the food and wood sectors by the duc de Coigny. Whatever their individual favourite domains and interests, the practices of these courtiers were broadly the same.

Artisans and court society

The *hôtel* Kinsky was undergoing constant alterations from the end of the 1770s until the death of its owner in 1794, and Thiéry's revealing portrayal in 1786 described it room by room, with the delight of a connoisseur:

> Crossing the winter dining room ... one reaches the magnificent music salon, decorated with fluted and gilded Ionic pilasters ... M. Julien de Toulon, Painter, has portrayed in the ceiling all the Gods assembled on Olympus to hear the songs of Anacreon, conducted by Apollo and the Muses.[15]

These pilasters, mirrors, figures, candelabras, niches, statues, gods, and copses were the work of several dozen artisans. Over a period of 30 years, the princess patronized more than 100 building tradesmen, including 17 painters and gilders, 14 florists and gardeners, as many sculptors and monumental masons, 7 upholsterers, 6 carpenters, and 3 cabinetmakers. The princess's behaviour and the vast sums she poured into her property, and especially her garden, were inseparable from the infatuation of French elites at that time with England and the Orient. The impact of Anglomania and Orientalism on landscape architecture was particularly strong, along with the vogue for English garden designs. Ruins, mountains, rocks, caves, rivers, islands, bridges, pavilions, varied walks, winding paths and charming sites were all features now deemed indispensable for those who prided themselves on their good taste.[16] Quite apart from the way they renewed the art of gardening, they became significant symbols in the social display of rank. Consider the Chinese pavilion built in 1773 at Chanteloup by the duc de Choiseul, Ermenonville Park dotted with the inventions of the Marquis de Girardin between 1766 and 1776, the follies of the Monceau Park created between 1773 and 1778 by the duc de Chartres, or Marie-Antoinette's Trianon hamlet, designed between 1783 and 1787. Following the fashion, Princess Kinsky appointed John Williams, gardener of the English nursery at Sèvres, to transform a classically designed French garden into an Anglo-Chinese Indian park, with a mountain and little pavilions, rocks, a lake and a hothouse filled with exotic plants. Two years later he added another mountain, a river, grotto, Chinese bridge and greenhouse.[17] The cost matched the ambitions of the project, reaching almost 15 000 *livres*,[18] without taking into account the payments to painters, carpenters, ironsmiths and masons for their work on the lakes, greenhouses and orangerie, and for the structures perched on the 'mountain' created at the bottom of the garden. For their work on the Chinese pavilion, built by the architect Bénard in 1790, the painters and decorators Moench and Langlois were paid about 4000 *livres*, the painter-gilder Montigny 1000 *livres*, the monumental mason Corbel 3000 *livres*, the cabinetmaker Jacob 2000 *livres*, and so on. The total sum for the alterations to the *hôtel* and the park in the years 1789-92 amounted to more than 160 000 *livres*. The ever growing number of tradesmen becomes more readily understandable in the face of this extravagant expenditure.

At least 50 artisans were needed for the constructions because of the highly specialized nature of the work.[19] Indeed, the generic term 'painter' on its own is rarely to be found in the craftsmen's bills which indicate the diversity of the skills supplied: décor-painter and architect, painter-carver and gilder, painter and decorator, house-painter and gilder, frame-painter and gilder, pattern-painter and gilder, decoration-painter and gilder, painter-gilder and building contractor, painter-gilder and varnisher, painter and stucco-worker. The statutes of the various corporations distinguish between colourmen, paint suppliers, metal gilders, those responsible for gilding copper, iron, cast iron or brass, stucco artists responsible for stucco ornaments, and painters who subdivide into house painters, gilder-painters, decorators, and members of the Academy of Saint-Luc, for only the greatest artists were freed from the obligations to the Corporation by their affiliation to the Royal Academy of painting and sculpture. Far from working in sequence, the painters employed by the princess worked simultaneously on different tasks.[20] Gibelin, the house-painter, Langlois, the décor-painter, Gouthière, the carver-gilder and painter and Watin, the painter-gilder-varnisher-colourman all held quite distinct skills: the first painted the façade of the *hôtel* or the walls of the dining-room, the second painted the eight sides of the little pavilion in Gothic style as well as the 'décor in the passage, hung with fringed twill drapes, which leads to the garden', the third decorated two pairs of candelabra and a fireplace-screen, the fourth supplied the materials, paint, plaster, oil, varnish, brushes, and so on.

An artisan's speciality rather than his proximity was therefore the main criterion when the princess made her choice: two-thirds of her tradesmen in the housing sector came from outside the faubourg Saint-Germain (Figure 5.3). She found her tilers and pavers in the faubourg Saint-Marcel, Wibert, in the rue des Postes near the Estrapade, Périac, Fontaine and Sibire in the rue neuve Saint-Médard; Régnault, the florist to Monseigneur le Dauphin at Roule; Drevault, garden designer, at the sand-pit of Vaugirard; Deleuze, painter and stucco-worker, in the rue de la Pépinière in the faubourg Saint-Honoré; Muller, stovemaker, painter and fireplace-gilder, in the rue de la Roquette beyond the porte Saint-Antoine, at the completely opposite side of town from her home. This spread of artisans in trades related to the house from as far away as the suburbs can be explained by their need for space: all the gardeners and florists lived beyond the city boundaries because of their nursery beds, just as numerous carpenters, sculptors and masons did so because of the accumulations of tools, timber, blocks of earth and marble, furniture,

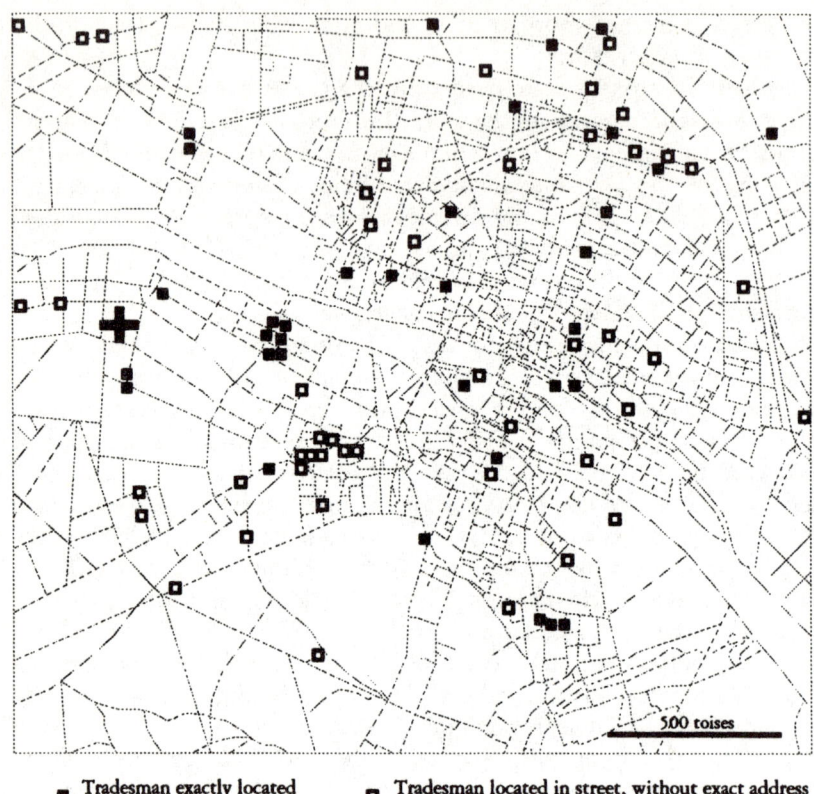

■ Tradesman exactly located   □ Tradesman located in street, without exact address
✚ Hôtel

Network of streets in Paris in 1780 following the 'plan des paroisses de Paris' drawn up by Junié, ingénieur-géographe.
Produced by Françoise Vergnault, July 1986.

5.3 The tradesmen of Princess Kinsky, rue Saint-Dominique (1773–94), housing trades

statues, paving stones, mouldings and cornices in their workshops. The engravings in the *Encyclopédie* show just how much space was needed for the cutting, sawing and assembling of wood, or for the cutting and polishing of stone. The distant locations of trades such as refiners and cesspit cleaners were dictated by the foul smells associated with their activities. At Lecolant's feather- and down-cleaning works, he carded and cleaned mattresses, bleached cloth, degreased 'wool and horsehair and removed all worms'. His establishment was situated in the rue du Faubourg-Saint-Lazare, beyond the Saint-Laurent fair.

It was because the house meant so much to her that Princess Kinsky sought her many artisans, those she judged the best in their trade, wherever in the city they were to be found. On the other hand, she found the trades related to horses much closer at hand, for here was a sector that concerned her far less: nine out of 10 tradesmen in this field lived in her *quartier*. Although these trades were particularly well represented in the faubourg Saint-Germain (Table 5.1), their density on the left bank cannot in itself explain the princess's choice. About three-quarters of the horse-dealers used by the Coigny family were located on the right bank where the family lived, indicating their strong presence on that side of the Seine. Choice was therefore not necessarily determined by the specific location of artisans in the city: it was also influenced by the location of the *hôtel* itself and above all by the strength of the family's interest in a particular sphere of expenditure.

The wide dispersal of the traders connected to horses who supplied the La Trémoille family confirms this point. In less than 20 years, the family used 115 artisans: at least 30 fodder merchants, 24 saddlers, 15 coach- and horse-hirers, 12 horse-dealers, 8 ironsmiths, 6 wheelwrights and 6 harness-makers. These high figures help explain the extensive distribution and result from both the large number of horses and vehicles used, whether hired or bought, and from the family's great commitment to the equestrian field which they regarded as an important investment, in just the same way as the house was for Princess Kinsky. The duke and his wife had at their disposal the widest possible range of carriages: town berlin and stagecoach, country berlin, cabriolet, barouche, dogcart, *vis-à-vis*, garden carriage, postchaise, sedan-chair. Their number largely explains that of the horses.[21] In 1782, when the duke announced the sale of the family residence in the Marais, the advertisement revealed that the small *hôtel* contained 'stables for 40 horses, sheds for 12 to 15 vehicles ... accommodation for horsemen, grooms and stable staff'.[22]

One finds an infatuation with English fashions here too, imported during the 1770s by a handful of young aristocrats, including several royal princes, all convinced anglophiles: the Counts of Lauragais and Artois, the Dukes of Lauzun and of Chartres, the Dukes of Coigny and of Fitz-James. England, the home of equestrian sports, became the model for the rest of the world. The races were ostentatious spectacles which henceforth found a firm place in elite sport and entertainment. Horse-riding as a leisure activity became indissolubly linked to the drive for luxury and distinction and, like similar 'objects', concerned above all with refinement. The racehorse was no more than a fleeting part of the show, for the saddle-horses and coach-horses were no less distinctively English. As Montesquieu observed, 'they say that the leader of Parisian society is the man who has the best horses to his carriage'.[23]

A good part of the La Trémoille stables were of English origin for the same reasons; hence the places of sale – Caen, Neuilly, Chaillot and London – as well as the prices paid for the horses which were much higher than usual,[24] reaching 1000 or even 1500 *livres*. The fashion for things English extended beyond matters of equipment (English carriages, saddles, harnesses, race-whips, and so on), to those of staff. The duke recruited an English jockey, groom and postilion. The equipment was an external sign of the master's wealth and opulence, but it also indicated the excellence of his taste. In their equipment and maintenance, the stables constituted one dimension of aristocratic consumption. The number of horses and coaches reflected the demands of court life and allowed the courtier to maintain a continuing display of status as he rode through the town and through the realm.[25] Keeping in fashion required a continual stream of new purchases: the duc de La Trémoille bought 15 carriages between 1777 and 1787 and more than a score of horses between 1777 and 1782, while his son purchased 16 between 1781 and 1785. The expenditure was inevitably considerable, almost 30 000 *livres* a year for the La Trémoille family.[26] A wide range of skills needed to be at their disposal and, just as with the princess, they were simultaneously calling on the services of several artisans from the same guild and changing them frequently: two-thirds of the saddlers and half of the ironsmiths figure only once. The same was true of the frequent succession of people working in the stables: coachmen, postilions, grooms, stable porters, huntsmen, equerry, stud-groom and jockey.

The role of the corporation was important, just as we saw with respect to the house sector. Saddlers and wheelwrights had had responsibility for the manufacture and maintenance of coaches before the explosion in the number of vehicles, but with more than 15 000 carriages in Paris in 1720, other corporations were permitted to compete with the saddlers.[27] Wheelwrights now made the wheels, blacksmiths and ironsmiths forged the axles, springs and hinges, harness-makers dressed the leather and joiners built the body, while saddlers adorned the interior and painters embellished the exterior. The saddlers' bills often referred to the work of ironsmiths, joiners, trimming-makers, mirror-cutters, wheelwrights and painters:

> the joiner fitted the edge of the moulding round the bodywork ...; removed the four mirrors from their frames to allow new ones to be fitted ...; completely stripped the front tambour so that the wheelwright could fit a new one; the gilder provided eighteen feet of beading ...; stripped the top of the door so that the locksmith could fit a lock.[28]

If we look at the saddlers, who were by far the most numerous artisans for the La Trémoille family in exactly the same way as were painters for the Princess, we find the same pattern whereby a single title hid a range of different occupational skills. Nor is this the only term used. A multiplicity of titles indicates subtle differences in the practice of the trade, and reminds us that the dividing lines between neighbouring corporations were not clearly drawn. The records of work yield saddler-harness-maker, saddler-wheelwright, saddler-painter, saddler-painter-wheelwright, saddler-ironmonger, saddler-coachmaker, saddler-coachmaker and horse-dealer, saddler-horse-hirer, saddler-carriage-hirer, saddler-fodder merchant. Taking the saddlers together, we find that three-quarters were carriagemakers and dealers, and that each of these appears only once. Some of the other six were in highly specialized trades, while others were in activities other than the sale of carriages, with its high dependence on fashion. As a result, their involvement was longer term. One of them rented sheds, needed because of inadequate space in the *hôtel*, while another rented out stagecoaches and funeral berlins and was loyally called upon at each bereavement, in 1773, 1779 and 1789. Three others had taken on annual maintenance contracts, and a last confined himself to providing equipment and materials, such as saddles, spurs, stirrups, whips and upholstery.

Similar causes produce similar effects: the distribution area of the 43 tradesmen patronized by the La Trémoille family covers a large span of the town (Figure 5.4). The duc de La Trémoille, like Princess Kinsky, chose his artisans for their skill rather than their proximity. A majority – 20 in all – were to be found in the faubourg Saint-Germain (of whom seven in the rue du Bac and five in the rue de Grenelle), but there were 11 in the north-east of Paris, 10 in the north-west and two in the south-east. The duke clearly did not restrict himself to his own *quartier*. The most striking feature of the distribution is the absence of the old centre of Paris: trades connected with horses need a great deal of space for sheds, workshops, warehouses, storehouses for fodder, carriages, animals and equipment. The closest artisans on the right bank were situated in the rues des Francs-Bourgeois, Michel-le-Comte, des Deux-Portes-Saint-Sauveur and Croix-des-Petits-Champs; on the left bank, a carriage-dealer lived in rue Guénégaud and a coach-hirer in rue Mazarine, a horse-dealer and another hirer were close to Place Maubert. The latter were well located, for from the time of Henri IV until its move to the Barrière d'Enfer in 1868, the horse market was held on the Boulevard de l'Hôpital, opposite the Salpêtrière Hospital, on the far eastern edge of the Notre-Dame *quartier*. An unfinished painting by Swebach-Desfontaines, *Le Marché aux chevaux*[29] (the horse market), depicts the horse-dealers pacing up and down before the horses gathered on the tree-lined boulevard, pausing now and then to examine their feet or mouths; in the background stand the pediments at the foot of the house built by Sartine in 1760 to house the market's director.

For many artisans, there is no apparent link between their occupation and their location in Paris. The 14 saddlers counted between 1781 and

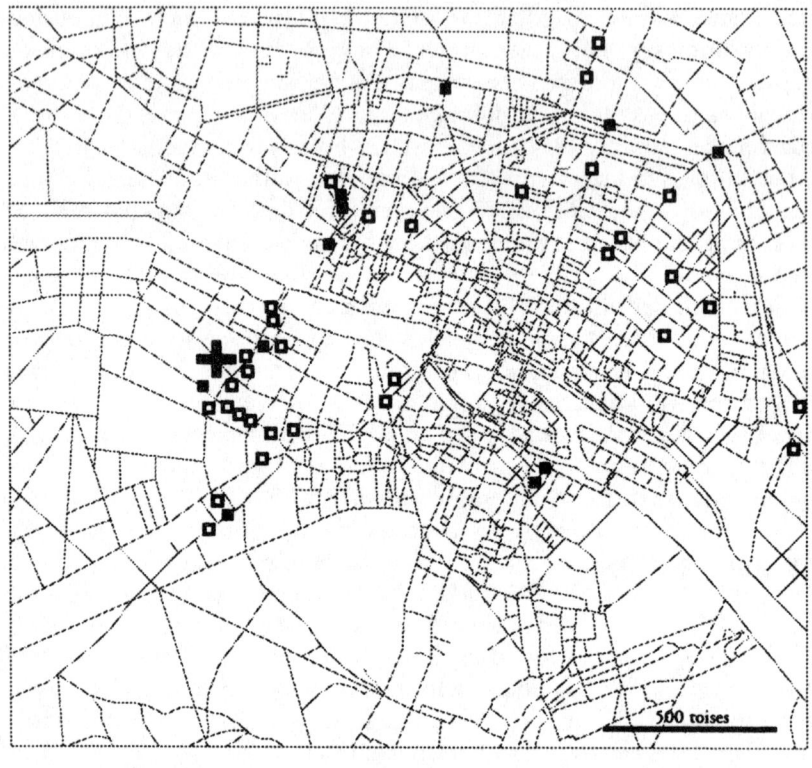

■ Tradesman exactly located ☐ Tradesman located in street, without exact address
✚ Hôtel

Network of streets in Paris in 1780 following the 'plan des paroisses de Paris' drawn up by Junié, ingénieur-géographe. Produced by Françoise Vergneault, July 1986.

5.4 The tradesmen of the La Trémoille family, rue Saint-Dominique (1781-92), horses sector

1792 were widely dispersed: half of them lived in the south-west; the others were scattered from the rue du Carrousel, near the Grandes Écuries, to the rue du Temple; the three artisans who maintained the vehicles and equipment were often needed on more than 10 occasions in a month, yet they did not live close to the *hôtel*, but in the rue Croix-des-Petits-Champs, rue du Temple and rue des Vieilles-Tuileries; the two artisans who made and sold harnesses lived near the Palais-Royal, while the artisans who repaired them were in the rue des Deux-Portes-Saint-Sauveur (Saints-Innocents) and rue du Faubourg-Saint-Antoine, while, to the west of Paris, one finds a horse-dealer in Chaillot and another in Neuilly. The picture is thus clear: the aristocratic residence drew its regular artisans and suppliers from the town as a whole, including its suburbs, and did not limit itself to the immediate neighbourhood.

The trading relationship and artisanal mobility

The distribution of artisans within the town raises the question of the way the process of exchange is organized, of the way in which the *hôtel* and the workshop are brought into contact. We need to understand exchange within its everyday context: the selection and transport of goods along with requests for services and their execution required a great deal of coming and going. We therefore need to grasp the mobility and movement that was inextricably bound up with economic activity. When the client was a nobleman, proximity was much less important since the journey was often undertaken by his servants or by the artisans and their apprentices who went to the *hôtel* to work there or to deliver goods. In this context the distance or proximity of the tradesman was of little significance for the act of purchase.

It was rarely the nobles themselves who made these journeys: their social position and their military or political obligations made it difficult, but it was excluded above all because economic activity was synonymous with the loss of noble title (*dérogeance*), which produced contempt for trading and artisanal activities.[30] They were therefore not normally expected to travel, and whether they in fact did so depended not on distance but either on their interest in the sector in question or on the status of the shop. The central issue was whether the shop was on the circuit of aristocratic sociability. In the case of an unattractive workshop-cum-warehouse, the noble would leave it to servants, messengers or the artisan himself to do the travelling. In the case of a fashionable salon-boutique, on the other hand, he may well have travelled so as to be there himself and to enjoy strutting about with his peers: shopping at fashionable establishments was not all that different from the well known practice of visiting, which gave the aristocratic day its structure. 'It was the great fashion to make morning visits. We all rushed to each others' places, certain to meet no-one at all, but it was all the rage.'[31]

The domestic servants were essential for all of this, organizing daily life in the most minute detail, in order to free the courtier from responsibility for any practical matters. The division of labour was precise. The butler, groom and housekeeper were in charge of the general provisioning of the *hôtel* - food and drink, fodder and household linen; the valet and chambermaid were the masters' messengers responsible for their personal purchases; the steward, who administered the property and business affairs, supervised all expenditure; the architect was responsible for everything related to his own sphere. Moreover, the servants were supported by a range of outsiders such as messengers, carriers and carters, whose work was to carry letters, messages or goods and who were used on a daily basis by both tradesmen and domestic staff.

Proximity was made even less significant by the fact that artisans would readily go to each other's workshops as well as directly to the *hôtel*. Indeed, the corporate system required a minutely regulated division of labour between different guilds before the product was finally completed. The chain of production thus, by its very nature, produced a good deal of movement which impinged on the process of exchange. As a result of these divisions many operations did not go through the *hôtel* at all: tradesmen would simply hand tasks on to other artisans who would actually go to the noble's residence. The timber merchant supplied the carpenter and the joiner who paid on delivery of the materials and were subsequently reimbursed by the steward; the joiner supplied the sculptor and the saddler; the cloth merchant provided materials for the upholsterer, the tailor and the saddler; the iron merchant supplied the wrought-iron worker, and so on. Bills occasionally mentioned such operations: '... lengths of moulding supplied by joiner ...'[32] or 'supplied to M. Denayve, tailor, three *aunes* [ells] of serge in blue silk'.[33]

The building trades, on the other hand, necessarily had to work on the premises, whether fitting a stove, mirror or curtains, repairing a chandelier or putting up wallpaper. An engraving by Roubo, *Manière de poser la menuiserie* (Methods of fitting joinery), shows four craftsmen at work in an antechamber, and depicts workbench, ladder, panelling and doors, for the joiners have brought everything with them.[34] Neither workshops nor shops expected to attract customers who knew in advance the product or service which they were looking for. The courtiers were not going to cross Paris to gather in some warehouse on the boulevards on the north side of the river any more than they would go into the eastern suburbs whose sad reputation was evinced by Mercier: 'That is the *quartier* where the Parisian populace are to be found living – the most poor, the most restless, and the most uncontrollable'.[35]

The workshop was functional and unattractive in appearance. Far from offering a brilliant display to its customers, it was generally a large and noisy shed where master and workmen banged, sawed, planed and hammered. Sculptors' workshops were, as a result, mostly set away from the main streets, looking in on a court or sometimes in a shed built within the courtyard of a house. *Le Ménage du menuisier*[36] (The joiner's household), painted by Lépicié, reveals to us not only the model virtuous artisan who was an industrious and caring father, but also a modest interior that served both as workshop and warehouse, and which was divided in two by a long bench over which the joiner himself can be seen bending; next to the fireplace stand laths and planks of different dimensions, and others are strewn on the floor, while in the background heavy panels and a large slanting beam can be seen. He may have rarely been visited by his noble customers, but the skill and reputation of the guild master were nonetheless important: the gap separating a tiler from the rue Gracieuse and a marble-mason of the royal buildings in the rue du Bac was a wide one. This same differentiation could be found within a single guild, for example between joiners and cabinetmakers, whose greater specialization in the making of luxurious and decorative furniture was far more prestigious. But even when an artisan was close to being an artist – as was a sculptor, monumental mason, carver and gilder or joiner and cabinetmaker – the workshop rarely served as a place of aristocratic sociability, in contrast to the fashionable boutiques or luxury ironmongers of the Palais-Royal and rue Saint-Honoré. The craftsman's trade, unlike that of the retailer, remained irredeemably tainted by low status. The artisanal sector of Paris could boast no equivalent to the Palais-Royal, where the elite of the corporate trades might have gathered close to their customers. The Louvre Arcades welcomed only painters and sculptors. Only a few outstandingly creative artisans could entice noble customers to leave the well trodden paths of the Palais-Royal and visit their workshops. When Baroness d'Oberkirch went to the faubourg Saint-Jacques, accompanied by the Duchesse de Bourbon, one would imagine that she was going to the farthest ends of the earth: 'We went as far as the faubourg Saint-Jacques, to a marvellous fan-maker called Méré. He was living *in a slum*; it is certain that neither Boucher nor Watteau have produced anything to match his gouache paintings.'[37]

Debt and the art of aristocratic consumption

The luxury of the court drew nobles into fierce competition for prestige, leading to unlimited expenditure and increasing debt, which was an unavoidable consequence of their position in court society. To be a courtier meant to spend, and the obligation was accepted without question. The eighteenth century retained vestiges of older noble ideals of honour and glory, giving without counting the cost and without counting the consequences. As a result, those so favoured by fortune that money flowed in from all directions were those who frenetically squandered it. The budgets of these high-ranking families were unbalanced even to the untutored eye, and with the courtier's lifestyle, generally speaking, went a chronic and increasing level of debt. The accounts of the duc de Coigny between 1757 and 1777 show a regular deficit: expenditure exceeded income every year by an average of 80 000 *livres*.[38] At the end of the 1780s, the Duke and Duchess de la Trémoille owed roughly 500 000 *livres* to some 40 individuals, while the debts of the Duke and Duchess de Fitz-James had by 1785 reached 800 000 *livres*. In the absence of savings, it was the *rente* – that is, the payment of interest on debts – which became the main means by which nobles could live on credit, and they did so without restraint. In 1789, the annual interest payments due from the Fitz-James family amounted to more than 90 000 *livres*. However, the *rente* represented only the visible part of the aristocrat's debt: servants, family tradesmen and unsecured creditors would not even benefit from this system of deferred payment, unless they chose to embark on long legal proceedings. The accounts occasionally reveal the extent of the money which such people were owed. Two-thirds of the 600 000 *livres*' expenditure of the Duke and Duchess de Fitz-James between 1786 and 1788 represented settlements of debts to more than 200 artisans and merchants. Although the largest payments were to cloth and fashion merchants, there were other substantial sums, such as 20 000 *livres* to a joiner, 17 000 *livres* to a jeweller, 10 000 *livres* to an ironsmith, 9000 *livres* to a carpenter, 7000 *livres* to a wheelwright and to a mirror-dealer, and 6000 *livres* to an upholsterer.

Tradesmen and artisans sought to protect themselves against the permanent financial difficulties of the courtiers whom they served. Economic reality threatened them with bankruptcy should their accounts fail to balance, but they were often able to protect themselves only at the price of endless legal proceedings. The game between court artisans and their aristocratic customers was a subtle one. The sheer range of their expenditure meant that nobles provided a living for a large number of suppliers who were more than ready to profit from their extravagance. These artisans may have been the masters of taste, but the immense sums involved and the fecklessness of their clients meant that any success they achieved served only to increase their dependence. The luxury trades had to submit to being ruled by credit. Credit was used in all sections of society, but nowhere was it stronger than amongst courtiers, because of their high incomes and enormous expenditure on the one hand, and their privileges on the other. If nobles occasionally paid in cash or with a delay of no more than a few months, they mostly paid by instalments or deferred settlement. One result was a staggering rise in the levels of their expenditure and indebtedness. The 'deferred economy' between client and tradesman which Daniel Roche has described in the clothing trades applied equally to other types of commodities.[39] Purchases were often paid for only after a considerable delay, which resulted in the astronomical sums reported above. A delay of 10 years or longer was by no means unusual. Tradesmen simply integrated this constraint within the way they ran their business, whether willingly or not, and only a minority insisted on being paid in ready money.

The sources reveal a war of attrition between the two sides, with pressure from the creditor answered with defensive action on the part of the debtor. The tradesman was less favourably placed, for the noble was a desirable customer, in principle at least: any confrontation meant the risk of losing his custom altogether. Moreover, it was difficult to stand up to a man of high rank in a society which rested on respect and deference to hierarchy. In the long run the tradesman generally obtained satisfaction, but he had to be firm and persistent. The stages of the process indicate methods with which we are now familiar: attempts to reach an amicable agreement, the search for compromise, threats of legal proceedings and, as a last resort, an appeal to the courts. The mounting number of formal accusations by tradesmen from the 1780s onwards shows the emergence of methods of resolving disputes over debt that are now the norm in contemporary capitalism. Confronted by his creditors, the aristocrat used every means available to gain time, resorting to a variety of subterfuges: ignoring the bill, quibbling about the price or even flatly denying it, disappearing altogether when demands were made, changing tradesman, invoking the king's justice. The tradesman had a variety of ways of his own in which to respond, by demands, entreaties, or resort to legal action. The tone of the tradesman's appeal to the nobleman varied with his temperament as well as his business situation. While some presented a litany of their miseries, lamenting their wretched situation – a blind son, an illness, or the death of a husband – others feigned disappointment or wrote increasingly menacing letters.

Only when all other means of persuasion had failed in the face of the nobleman's sheer inertia would the tradesman decide to begin legal proceedings. It was a last resort, to be employed only when the survival of the enterprise was at stake. It was said of a blacksmith in just such a situation:

> The courts were the only hope left to him, but he had a peaceful spirit and found such action offensive. He hesitated for a long time but eventually, seeing that he had no other means available, believed that he could somehow invoke the magistrates' judicial powers without injuring M. le duc de la Trémoille.[40]

A variety of letters and legal documents confirm the extent to which such action was a last resort, and that the threat of bankruptcy alone forced the artisan's hand. Time was an economic imperative, and if the tradesman were to have a chance of success then it was something which he had to control. The nobleman, on the other hand, thought that he was free of such constraints, yet they were in reality unavoidable. The method was effective, for legal proceedings usually ended successfully, and, what is more, the law allowed interest to be charged on the sum due. Yet it all took a long time – three or four years was the least one could expect. The courtier would not accept defeat now that the dispute had moved on to judicial territory, however clearly in the wrong. Leaving aside the slowness of the legal proceedings themselves, nobles had not only resources but also privileges which they would shamelessly exploit. It is nevertheless interesting to observe that even the greatest noble families were not immune from the law: the Fitz-James family suffered at least 60 legal actions and the La Trémoille family about 40. They almost always lost: verdicts from the Châtelet and a range of summonses rained down on them, such as *assignations*, *commandements* and *commandements recordés*, plus distraints such as *saisie-arrêts* and *saisie-clôtures*. At the same time creditors kept a close eye on inheritance and the sale of debtors' property. The list of those objecting to the release for sale (*levée des scellés*) when furniture and goods were disposed of after death was often long: 10 in the case of Princess Kinsky, 17 for the marquise de Fleury (including 12 traders and artisans), 62 (including 46 traders) for the Maréchal de Fitz-James in 1787 (on account of his spendthrift son, the duke). At the end of the day the courtiers had to pay, whether in ready money, if an inheritance or the sale of an *hôtel* allowed that, or in longer-term *rentes* or assignation of stocks (*délégations* on the *Trésor royal*, the *Fermes*, or whatever). Luxury had its constraints.

The increasing number of disputes and legal conflicts during the period under study here, right up to the eve of the Revolution, suggest that a fundamental change was underway. Notwithstanding all the payments, pensions and legal support afforded by the monarchy – indeed these became chains as much as support – the financial problems of the aristocracy reveal the clash of opposing economic mentalities, those of court society and of the entrepreneurial bourgeoisie. The first remained driven by symbolic and cultural imperatives: expenditure for status and 'prestigious waste' prevailed, rather than the accounting concerns regarded as the preoccupation of *roturiers*, the common people. The second, whose social ascent was very much connected with this extravagant pattern of conspicuous consumption, had their capital invested in the market, and they responded to imperatives of profitability which were incompatible with the aristocrat's more casual attitude to money. The tradesman had to match expenditure to income if his social existence was to be maintained. The social existence of the great lord, on the other hand, could be maintained only by matching his expenditure to the expectations that went with his rank. This fundamental contradiction, aggravated by the *ancien régime*'s chronic shortage of money, was probably one of the reasons for its collapse.

Notes

1. J. Savary des Bruslons, *Dictionnaire universel de commerce, contenant tout ce qui concerne le commerce qui se fait dans les quatre parties du monde* ..., Paris: J. Estienne (1723-30), vol. I.
2. R. Mousnier, *Paris capitale, au temps de Richelieu et de Mazarin*, Paris: A. Pédone (1978), p. 163.
3. D. Roche, *La culture des apparences. Une histoire du vêtement (XVIIe-XVIIIe siècle)*, Paris: Fayard (1989), p. 54.
4. N. Elias, *The Court Society*, Oxford: Basil Blackwell (1983), p. 270.
5. G. Chaussinand-Nogaret, *The French Nobility in the Eighteenth Century. From Feudalism to Enlightenment*, Cambridge: Cambridge University Press (1985), p. 53.
6. The granddaughter of Count Jean Palssy, palatine of Hungary, famous for having repelled the Turks before Belgrade in 1717, Marie-Léopoldine-Monique, Countess Palssy d'Erdod (1729-94), had, in 1748, at the age of 19, married François-Joseph, Prince Kinsky. The young lady became a dowager in 1752. In 1766, she obtained naturalization papers. Princess Kinsky is by origin a special case, particularly since, in spite of her efforts, she did not manage to gain an introduction into the highest court circles.
7. The men rode in the king's coach and followed him at the hunt; the women were presented ceremoniously to the king, queen and royal family.
8. Between 1768 and 1784.
9. Between 1780 and 1787.
10. The exact total was 1789 tradesmen, of whom the address of 1299 is known: 1116 in Paris, 72 in Versailles, 102 in the Paris region or in the provinces, 9 in England. The majority of the material in the remainder of this essay is drawn from Natacha Coquery, 'De l'*hôtel* aristocratique aux ministères: habitat, mouvement, espace à Paris au XVIIIe siècle', doctoral thesis, University of Paris I, 1995. See esp. Part 1, 'L'*hôtel*, lieu et modèle de consommation', pp. 15-219.
11. Thus, the food sector includes butchers, bakers, candlemakers, confectioners, mineral-water dealers, wine and wood merchants, roast-meat vendors, vinegar-makers, etc.; the horse sector includes grooms, coachmakers, wheelwrights, equipment-joiners, saddlers; the habitat sector includes architects, carpenters, florists, masons, joiners, painters, plumbers, sculptors, glaziers; the hygiene sector includes apothecaries, surgeons, hairdressers, perfumers, wigmakers; the clothes sector includes laundresses, hosiers, bootmakers, embroiderers, drapers, braid-makers, stocking-makers, cloth merchants, milliners, furriers, ribbon-makers, tailors; the luxury sector includes pyrotechnicians, gilders, fan-makers, printers, jewellers, goldsmiths, picture-painters, porcelain-manufacturers.
12. Each merchant was recorded by street, sometimes with the precise location in the street specified, for example 'in the corner of street ...', 'opposite street ...', 'opposite the water-pump', 'alongside the church', 'at the foot of

the bridge', 'near the market'.
13. Fifty-nine per cent of the streets had just one merchant in 'luxury', 61 per cent 'horse', 64 per cent in 'clothes', 68 per cent in 'hygiene', 69 per cent in 'habitat', 72 per cent in 'food'; most of the streets containing between one and five shops cover between 93 per cent and 97 per cent by sector.
14. D. Roche, *La culture des apparences*.
15. L.-V. Thiéry, *Guide des amateurs...*, Paris: Hardouin et Gattey (1786-87), vol. I, pp. 581-7.
16. From L.-V. Thiéry's description of English landscape gardens in the capital, *Le Voyageur à Paris...*, Paris: Gattey (1789), vol. II, pp. 52-8.
17. Archives Nationales (A.N.): T 220/5-7: *Mémoire de ce que pourront couter les changements à faire au jardin de Madame la Princesse de Kinski compris la rivière, les rochers, la grotte, le pont et chassis de la serre chaude et entretien du jardin pendant trois mois ...*
18. Between 1779 and 1792, flowers and plants cost 7500 *livres*, plus 7000 *livres* for work done and 1200 *livres* a year for upkeep; the known total is more than 31 000 *livres*, an average of almost 2500 *livres* a year.
19. Four masons, cabinetmakers and monumental masons (four for each corporation); three sculptors and stove-dealers; two joiners, carpenters, ironsmiths, painter-decorators, mirror-workers, glaziers, plumbers and tinsmiths; a roofer, painter, gilder, carver and gilder, caster and fountain-maker, tiler, paver, lattice-worker, coppersmith, stucco-worker, gardener (plantations and terraces), a worker in charge of the 'state of the house', with the architect: a total of 48 in all: *État des honoraires dus par Madame la Princesse de Kinsky, à Charles Joachim Bénard, architecte; pour projets, détails, conduite, vérifications et règlemens d'ouvrages de bâtiments, dans le cours des années 1789, 1790, 1791, et neuf premiers mois de 1792 ...,* A.N. T 220/2.
20. In the garden, Montigny was responsible for the hothouse, orangery and Chinese pavilion, while Moench and then Langlois worked on the Indian pavilion; Ménagé was employed as an expert to check another artisan's accounts, while a sixth person took responsibility for an *hôtel* in the rue de Richelieu.
21. Two outings in a day by the lord and lady could exhaust at least 15 horses.
22. A.N. T 1051/48-9.
23. Letter 88 from Usbek to Rhédi, in Montesquieu, *Persian Letters*, Harmondsworth: Penguin (1993), p. 169.
24. An average horse cost 300 *livres* in the middle of the century.
25. Cf. J. Duma, 'Le comte de Toulouse et le duc de Penthièvre (1678-1793). Étude d'une nébuleuse aristocratique', doctoral thesis, University of Paris I, 1993, p. 708; see esp. ch. 16: 'Le système des dépenses: les formes de la consommation, transporter le Grand', pp. 671-708.
26. Some 6-8000 *livres* of this was spent on fodder alone. The receipts of the oat merchants to the duc de la Trémoille reached an average of over 3000 *livres* a year for 150 *setiers* (measures of 150-300 litres) of oats; those from oat, straw and hay merchants amounted to over 5000 *livres* for roughly 6000 bundles and about 100 *setiers*.
27. A. Franklin, *Dictionnaire historique des arts, métiers et professions exercés dans Paris depuis le XIIIe siècle*, Marseille: Laffitte (1977 edn), article on 'carrossiers'.
28. A.N. T 1051/42-3: *Mémoire des fournitures faites pour l'entretien des*

équipages de S. A. Monseigneur le duc de la Trémoille ordonnées à Michau sellier, 1770.
29. Musée d'Art et d'Histoire, Metz.
30. 'The artisans of towns have the same lowness of spirit [as one finds amongst country dwellers]', d'Alembert and Diderot, *Encyclopédie...*, Stuttgart-Bad Cannstatt: Friedrich Frommann Verlag (1966, originally 1751-80), vol. 2, article 'Luxe', p. 710.
31. Baronne d'Oberkirch, *Mémoires...*, Paris: Mercure de France (1989, originally 1853), p. 307. The *Mémoires* were written in 1789.
32. *A.N.* T 1051/42-3: *Mémoire des fournitures ...*
33. *A.N.* T 1051/44-5: cloth merchant's account to the duc de la Trémoille, October 1778-February 1780.
34. A.-J. Roubo in B. Pons, *De Paris à Versailles 1699-1736. Les sculpteurs ornemanistes parisiens et l'art décoratif des Bâtiments du roi*, Strasbourg: association des publications près les universités de Strasbourg (1986), pl. 2.
35. L.-S. Mercier, *Paris le jour, Paris la nuit...*, Paris: Robert Laffont (1990), p. 72.
36. Musée des Beaux-Arts d'Orléans.
37. Baronne d'Oberkirch, *Mémoires*, p. 447. Emphasis added.
38. At that time, the duke's annual income reached an average of 100 000 *livres*.
39. Roche, *La culture des apparences*, p. 201.
40. *A.N.* T 1051/50-55: written in 1784 in connection with Boutbart, blacksmith, of Attichy.

CHAPTER SIX

Craftsmen and revolution in Bordeaux*

Josette Pontet

In attacking the foundations of *ancien régime* society, the French Revolution could not leave unchanged the traditional organization of labour based on *corporations*,[1] privileges and monopolies. Guilds were part and parcel of the feudal system and on the night of 3 August 1789 the Club Breton had passed a motion asking for 'the abolition of all the privileges of class, province, town and guild'.[2] A world composed of communities defined by the privileges which the king had granted them was to be replaced by a new society based on equality and free access to all offices. However, bringing down the whole guild system was not an easy thing to do, as Turgot's unhappy experiment had shown in 1776.[3] Consequently the declaration passed on 4 August called for no more than 'the reformation of the trades'. The labouring world and its traditional organization remained beyond the scope of the legislation passed on 5-11 August 1789. Guilds were abolished only on 2 March 1791 by the *loi d'Allarde* that proclaimed free contract to be the foundation of the new social order and abolished all bodies and regulation that hindered the production system. Freedom of enterprise and freedom of work received official approval and three months later the *loi Le Chapelier* prohibited any kind of trade organization as well as combinations.[4] But on 26 August 1789, the Declaration of the Rights of Man proclaimed all men free and equal in rights and equality as the new guiding principle of society. Although guilds were still not abolished, many journeymen could claim the right to become self-employed and such a period of uncertainty and ambiguity could lead to tensions between masters and journeymen. It is a valuable exercise to study in a general way the changes induced by the French Revolution in the labouring world of a substantial town such as Bordeaux, numbering 110 000 inhabitants after a period of remarkable expansion,[5] and where the guild system, if not predominant, could enjoy some importance at a local level in defending the interests of the organized trades.

In no town did the guilds enjoy a total hegemony over the crafts, and

---

*Translated from the French by Philippe Chassaigne.

Bordeaux was certainly no exception. The guild system received there a rather late start with no such statute recorded before 1453, and it developed very slowly and in a relatively limited manner.[6] The difference with Bayonne was striking: the town was granted a charter in the middle ages and there is some evidence of organized trades in the twelfth century. If new trades developed later at a somewhat similar pace to Bordeaux during the second half of the fifteenth century,[7] their importance was in no way the same during the early modern period: with 14–15 000 inhabitants in the 1760s, Bayonne numbered 39 regulated trades and 843 masters,[8] while Bordeaux with a population five to six times greater numbered only 50 such trades and 1838 masters, that is, not much more than twice the number in Bayonne.[9] Moreover, the Bayonne trades were prevalent in the staple industries, while in Bordeaux they only existed in peripheral activities and there were none in cooperage, where the larger part of the local labour force was employed. This gives an idea of how important unregulated labour was in Bordeaux. A 1762 survey ordered by the Controller General of Finances recorded 38 free trades amounting to 2142 independent workers,[10] and yet this was clearly an underestimate, as demonstrated by the 1784 *Arts and Trades Almanac* which enumerated more than a thousand small free masters, a group which had gone unrecorded in the 1762 inquiry.[11] It is no exaggeration to say that the population of free artisans and merchants (i.e. those who were not regulated by a statute and only abided by the general rules, which does not mean of course that there was no supervision exercised by the town authorities) amounted to twice the number of guild masters. Almost all these trades were involved with commercial port activities, such as delivery trades (porters, rollers, watermen)[12] or trades linked to transport: shipwrights and coopers numbered no fewer than 766 in 1762 and 946 twenty years later, according to the poll tax rolls.

These trades had existed for a very long time, but demographic growth linked to economic expansion stimulated new commercial and productive activities that stood on the margins of the guild system.[13] Guilds were far from covering all the labouring classes, just as they were far from covering the city's geographical area. In the same way as in other towns, Bordeaux contained privileged places which were protected from the military authorities or the local stock exchange. Above all there were two *sauvetés*: the first and more important one was under the jurisdiction of the chapter of Saint-Seurin and covered about 20 sq. km in the north-western suburban area of the town, while the second and much smaller one (80 to 100 acres located within the walls[14]) was under the authority of the cathedral chapter of Saint-André. There were no guilds here and no masters: only a licence was needed to run a shop.[15] In

1777 there were no fewer than 115 tailors in Saint-André, and 50 in Saint-Seurin, while in the rest of Bordeaux there were only some two hundred masters. The town itself was naturally the main market for the craftsmen of the *sauvetés*, and especially those of Saint-André, whose boundaries with the town were in no way watertight. The guilds obstinately fought the illegal competition exercised by these unregulated workers whose legal status they sometimes refused to accept: guild bailiffs would on occasions search their premises and claimed, according to ancient statutes, the right to seize their goods.[16] Such efforts were very often in vain: their Bordeaux customers felt little sympathy for the masters' monopoly and showed solidarity with their suppliers in asking for the end of these practices.[17]

The multiplication of *chambrelans* was another severe threat to the monopoly of the guilds. The name says it all: these were self-employed people working in their own room without being masters or necessarily having any kind of qualification. There were many shoemakers, tailors and wigmakers among them, though every industry suffered from these false workers' (*faux ouvriers*) and all the corporate trades complained about them. Complaints, threats, direct action[18] or lawsuits were of little avail against their ever-increasing number in a town that was in full demographic expansion. These undercover workers were difficult to spot and, to an even greater degree than the workers from the *sauvetés*, received active sympathy from their fellow inhabitants who belonged to the same social group and cared very little about the guild police;[19] moreover, they provided them with goods at a much lower price than the masters of the official trades. Although they worked at home and consequently had no pure legal status, needlewomen and dressmakers enjoyed more toleration. Whereas in Paris they were organized in independent trades,[20] the needlewomen of Bordeaux (as ever, suspected of leading immoral lives[21]) had no authorization to exercise their trade. They were nevertheless quite numerous. In the case of dressmakers, widows or daughters of a master practising at home were affiliated to the guild of tailors; those who were totally independent, without legal recognition, enjoyed only very limited toleration.[22] Along with the very numerous seamstresses[23] they represented the bulk of the female workforce, after domestic servants.

The proliferation of *chambrelans* is very difficult to quantify, even if we may assume that they were numerous[24] as a consequence of a strong immigration of artisans and labourers into Bordeaux,[25] and it was an obvious problem for the guild masters. They found it increasingly difficult to enforce their monopoly over these 'false workers' and claimed to be their victims, in spite of the heavy fines imposed on them and the active support guild masters received from the town council or the local *parlement*, which still held the guild to be the best means of upholding law and order. Sometimes a monetary agreement could be reached between bailiff and *chambrelan*. In June 1788, such a tailor was left in peace after he had paid a fine of 60 *livres tournois*.[26] Yet it cannot be denied that these illicit activities spread and shook society to its very roots. Competition was introduced in the market, the prices dropped and the economic freedom advocated by local economists and physiocrats, such as the Abbot Sicard, a disciple of Turgot, Isaac de Bacalan or the people of the Museum,[27] was introduced into the daily life of the city. Representatives of the trades criticized the fact that some masters did not hesitate to give work to *chambrelans*, in violation of the most basic rules of the guilds.[28]

Indeed the system suffered blows both from the outside, since an increasing part of the labour force produced, sold and offered services outside the system, and from the inside. The guilds multiplied until the middle of the eighteenth century and their functioning was less and less coherent.[29] Both the regulations and the philosophy of the guild system, as fixed by various statutes, were openly questioned. Many masters became members without producing the masterpiece required by statute,[30] either because rural journeymen lacked proper training[31] and were less able or because they were exempted after the payment of substantial compensation. This seems to have been the case in all the metal trades, from boilermakers to ironmongers and gunmakers,[32] and can also be found among goldsmiths and wigmakers. Money was gradually becoming more valuable than knowledge. Such an evolution had undoubtedly been favoured by the monarchy, and the king was able to exempt those who sought to be masters from the requirement to have been indentured. None of the 25 goldsmiths who became masters between 1776 and 1789 had been through such a process.[33] This situation led to complaints and the monarchy undoubtedly adopted a very ambiguous attitude towards the guild system. It wanted at the same time both to reform abuses and to uphold the guild system, as shown by the edict of August 1776 altering that of Turgot passed in February the same year, which was concerned with the 'abolition of the *jurandes* and communities of commerce, arts and trade', and to sell royal letters that could make a master of someone without entry fees, or masterpiece or any training in the trade. However, these edicts did not have any effect on the Bordeaux guild system: the February edict was never implemented there, since the local *parlement* had not been consulted, and the guilds remained unaffected by all the Parisian turmoil surrounding them.[34] As far as the August edict is concerned, the *intendant* Dupré de Saint-Maur was in favour of economic freedom and prevented its application. Nonetheless the king, taking into consideration the requirements of the market (there were complaints about the scarcity of masters in some trades, such as wigmaking[35]), but also with purely financial considerations in mind, sold royal letters to private persons who would then be full members of the guild. Dupré de Saint-Maur was at first against such a procedure, but he ended up agreeing to a hundred or so requests in order to open up the trades and allow journeymen to become masters. It goes without saying that the integration of these newcomers was not achieved without tensions.[36] These practices had important consequences. Consulting a master was no longer a guarantee of quality. Moreover, the acquisition of a royal letter or an indenture often had a purely financial aim: letting at a profit was of course forbidden but widely practised. There was a real speculative market for it: a M. Lacouture acquired seven of these licences for wigmaking; he let out six of them for 160 *livres* a year, and each leaseholder would become an owner in his full right as soon as he had paid its estimated value of 3500 *livres*. No mastery of the trade was required in counterpart.[37] This was far from the original ideal of the guild system, even if licensed wigmakers had always been in the habit of disposing freely of their licence. Yet the possession of such a licence had become a way of making money. Mergers were commonplace and it was no exception to see a master running a couple of shops in spite of the official interdiction – this was especially the case among tailors.[38]

There was a fair amount of competition between masters because of an essential feature of the trade: customers' freedom of choice and the law of supply and demand meant that the masters were eventually placed on an unequal footing.[39] The geographical expansion of the town and the uneven distribution of the trades in the city[40] had the same effect: masters working in the rich aristocratic districts enjoyed a far greater commercial potential with rich, high-spending customers who would choose the best producers and traders. Besides, the distribution of shops in the city made it increasingly difficult for the police of the trades to enforce the existing regulations on production, workforce or trade. We can see everywhere,[41] and particularly in Bordeaux, a shift from production to commerce: for instance, many tailors were merchants as well and provided their customers with fabrics to the detriment of merchant drapers who claimed to have the monopoly.[42] In the same way that there were rich and poor trades,[43] there were rich and poor masters in every trade even if the proportion of the former varied in importance: 1 per cent of master tailors paid 50 *livres* in poll tax, but 22 per cent of butchers and 23 per cent of surgeons; to this elite we could add masters paying a poll tax of at least 10 *livres*: 20 per cent among tailors and 34 per cent among shoemakers; but half the tailors and half the shoemakers paid only 5 *livres*. The difference between them and their journeymen was small, though the latter were on the whole poorer since 90 per cent of them paid less than 5 *livres*.[44] Social contrasts were also perceptible among the butchers where only a minority could be labelled as 'well-off'.[45]

Such a hierarchy was not specific to the guilds: it could be found among all Bordeaux artisans, and especially among coopers who were the most numerous of them, and where there was a true elite comparable to lawyers, attorneys or other judicial sub-officers, although these seemed to carry a higher social status.[46] Between 1764 and 1788, seven coopers left at death properties worth more than 1000 *livres*, compared to nine judicial sub-officers. This percentage of 'rich' amounted to 10 per cent in each category while those close to destitution amounted to an equal percentage of each group.

If the social diversification of the trades was a reality, it should also be noted that the traditional production unit, bringing under the same roof master, journeyman and apprentice, once the cornerstone of the guild system, had largely disappeared. Fewer and fewer journeymen lived with their masters and many apprentices, especially among wigmakers who had the reputation of being somewhat unstable, failed to return home at night, all too often with the complicity of their masters.[47] It would of course be wrong to see in the past a social harmony that suddenly came to an end in the eighteenth century. Trade disputes were numerous and ancient,[48] but it is certain that relations within the trades had slowly polarized. Urban growth brought many immigrants who arrived more or less qualified. Bordeaux masters had few apprentices: the 1762 survey recorded only 152 of them for 2500 craftsmen, and Gallinato numbered only 2538 indentures for 93 trades, that is to say about one a year for each trade between 1743 and 1790.[49] The indentures often stipulated that the apprentice would be paid a wage, provided he lived elsewhere in the city. The private and trade spheres were becoming increasingly distinct, as can be seen in the specificity of rooms in the dwellings of richer artisans, be they coopers or master tailors. The new importance of furniture in addition to the indispensable beds showed a new interest in everyday domestic life and no doubt in intimacy too. Very few inventories after death mentioned the presence of mats in the garret or elsewhere in the house for apprentices to sleep on.[50]

The corporation no longer structured individual life, in the way in which the spirit of the institution seemed to require. Many journeymen theoretically recruited by the guild[51] worked on a daily basis and there could be no personal link between masters and workers, whose relations were more and more conceived as relations of inferiority of journeyman to master.[52] A great many journeymen had only limited prospects of ever becoming a master, and the condition of journeyman, which had above all constituted a transient situation between apprentice and master, became more and more a permanent status. The trades were not, however, in most cases closed to immigrants. If masters' sons still had priority, they were no longer sufficient to maintain recruitment in many trades. Among the 250 tailors who became masters between 1757 and 1783, there were only 16 masters' sons and 44 sons-in-law, that is to say fewer than one in four.[53] The ratio was even lower among wigmakers, where 80 per cent were foreign to the trade. This kind of mixing was not good for the guild system, which had been set up in a closed economic environment where personal connections were paramount. In any case, the dramatic growth in numbers as a consequence of the expansion of urban demand and royal policy was certainly not conducive to the rigorous enforcement of the guild system. The case of wigmakers is particularly telling. Three batches of royal licences in 1762, 1781 and 1786 boosted the number of master wigmakers from 74 to 200.[54] Such a large assembly proved impossible to summon. A first attempt on 25 June 1787 turned into a brawl in the face of the jeering and insults of young masters 'who respected neither the syndics nor the elders'.[55] The guild found only one way of escape, which was to allow fewer masters to attend these excessively large assemblies. Only 48 of them would be allowed to deliberate (30 'ancient' and 18 'young') while two young masters were allowed to attend the drafting of the accounts to avoid any suspicion.

The system in wigmaking was overwhelmed by the number of its members and the pressure from young masters who rejected the guild system and asked for new types of regulation. Such a cleavage could be found in many other trades and was undoubtedly one of the key elements in the crisis of the system.[56] This can be attributed partly to the royal policy of selling licences, which led to the swamping of the old guilds. It can be also attributed to urban growth which acted as a powerful force weakening the guilds. This seems to contradict what happened in Turin, where the guild system was organized very late and was contemporaneous with the great phase of urban expansion in the 1720s and 1730s.[57] Cerutti has shown clearly, however, how non-economic elements were important in the emergence of guilds and in the changes in the way the city was perceived. The corporation seems to have become the place where an increasingly uniform and homogenous population could find differentiation.[58] This is far from the situation in Bordeaux in the later years of the *ancien régime*. The Revolution took place in a labouring world in turmoil, where deregulation had been increasing in practice if not in law. The separation grew between the community of the trade and a section of its members whose interests seemed increasingly divergent.

Corporations continued as a legal framework and their ruling authorities kept on fighting the perceived threats of freedom of enterprise and open competition, while entrenching themselves behind the privileges and monopolies they sought to defend forever, at the very moment when the *ancien régime* was nearing its end. The town council and the *parlement* always supported them, even if their support did not prove very effective.[59] Guilds remained useful for tax assessment purposes or militia raising, to such an extent that unorganized trades were turned into temporary guilds, as for the election of the representatives to the Estates General:[60] some 135 'guilds' or communities took part in the election of 240 representatives of the Third Estate.[61] The Third Estate had a rather diverse view of them, but on the whole it favoured their abolition, while the corporation masters asked that they continue. The division between supporters of liberalism and those of interventionism was clear. In Bayonne, the Third Estate called in its *cahiers de doléances* (grievance lists) for the abolition of the guilds, which in turn raised unanimous protest from the 'masters of all the guilds' who addressed a petition to the king: 'Thou shall not suffer to be abolished the establishments which propagate and improve the useful arts, and exercise a police function in the various trades'.[62] Their value was constantly asserted. There is no evidence of such a petition in Bordeaux but we know that, as in many other towns, guilds gathered their forces and asked the king to protect them, and they decided in a general assembly to send a memorandum to Necker.[63] The corporations pleaded once again their commitment to the existing system and asked, as did buttonmakers or shoemakers, to be left with their rights and privileges.[64] Their acceptance at the same time of the need to reform existing abuses revealed the limits of their commitment to the ideas of 1789.

The period between August 1789 and March 1791 revealed clearly the cleavage between masters and journeymen: the former were in favour of the status quo, while the latter wanted to be freed from the constraints of the guilds. Tensions abounded. Wigmakers provide a good example: as early as the end of August 1789, many apprentices refused to abide by the regulations of their trade 'because they were abolished by the Declaration'. They thought they were free to become self-employed and they refused in any case to obey their masters any longer. This was a problem acutely felt in all cities. Throughout 1790 the municipal council of Bordeaux was pestered by the syndics who kept on asking for the statutes, which were still valid, to be strictly enforced, and wanted any encroachment to be severely punished, even by prison sentences. The council kept to its traditional policy, initially treating harshly journeymen who were supposedly misusing their newly acquired freedom, but the council elected in April 1790 seemed more divided. It judged that it was 'neither possible, nor appropriate, to be strict about all the regulations that had been established during the Ancien régime'.[65] The 'new state of things' which so worried the town council was to bring about the abolition of the guilds. The *loi d'Allarde* was passed by the Assembly on 2 March 1791 and enacted on 17 March. Yet the corporations had attempted as early as February 1791 to persuade the Constituent Assembly to perpetuate them, and had vainly tried to communicate between towns about the best approach to adopt.[66] Public opinion favoured the new ideals of freedom and individualism and was at best indifferent to the cause of the guilds. The corporations reacted little to the passing of the new law. Some of them, such as tailors, wigmakers and joiners, tried to carry on as before, but this could be no more than a rearguard action. Master wigmakers objected that they had a right of property in the mastership which they had obtained at a good price. Indeed the problem with the *loi d'Allarde* was in getting the state to compensate for the licences or the offices, since it had instituted a new tax, the *patente*, whose payment was obligatory for all who ran a shop on a self-employed basis, and the masters wanted at least to obtain a refund on their old licence before they complied with the new rules. This was not granted until 1793. Wigmakers complained repeatedly until then that trade had been ruined by both competition from their former apprentices and political events.[67]

The words *maître* and *garçon* disappeared from the vocabulary, just as the signs of the guilds drifted from the urban landscape, but the process was a slow one: those of tailors or tobacconists had to be removed in 1798 and in 1800 respectively.[68] Expectations were running high among journeymen about the abolition of the guilds, but they were disappointed when the *loi Le Chapelier* forbade any kind of combinations. Journeymen's *compagnonnages* were to some extent reconstituted undercover, and unrest continued among bakers, tailors and nailmakers.[69]

Did the new legislation alter the labouring world in practice? A few apprentices certainly took advantage of it to become self-employed, provided they paid the new *patente*, but it meant very little to unregistered *chambrelans*; merely that they would no longer be prosecuted by the police of the trades. Marriage settlements made in 1789–1815 reveal that some tailor- or wigmaker-assistants became self-employed,[70] but this was relatively exceptional. The *patente* was proportional to the rental value of the shop and its payment involved a certain amount of money, as did renting a shop or buying the basic equipment. In order to calm the worried wigmakers, the Society of the Friends of the Constitution stipulated in 1791 that 'the expenses and the uncertainty of starting a business would discourage many a worker, who will then prefer even a modest salary to uncertain profits that could cause his ruin'.[71] This turned out on the whole to be true. The condition of many journeymen remained unchanged other than in name. They could of course still marry a master's daughter whose wedding presents would include a shop and a profitable succession agreement.[72] On the whole apprentices were no more numerous in Bordeaux than they had been before.

It is not easy to judge with any accuracy whether the tailoring and fashion trades were particularly hit by the uncertain situation at the time. In the same way as their counterparts in Paris, Bordeaux's wigmakers diminished significantly in number, though less acutely.[73] While 203 masters had taken part in the 1789 assembly, there were only 171 registered in the 1791-92 almanacs, that is, a decrease of slightly less than 15 per cent. The decline was particularly noticeable among young masters: over a quarter of the new masters of 1786 had disappeared by 1791. It is possible that they had incurred debts in order to buy their licence, and both the difficulties caused by 1789 and the competition of their former apprentices (who were unrecorded in the 1791 almanac) proved to be fatal. Master tailors were less exposed to competition and seem to have fared better. The range of incomes within the trades had not lessened, but it was no longer possible to distinguish master from worker. The extent of poverty among tailors can be seen in the marriage contracts registered with a notary between 1789 and 1815: over half (53.7 per cent) of the partners had no personal property, and only 15 per cent of them had an annual income estimated at 50 *livres*. Less than 20 per cent had a property valued from 50 to 400 *livres* and 11 per cent disposed of more than 400 *livres*, the richest one boasting over 10000 *livres* worth of property.[74] Destitute tailors were a larger proportion of the trade than they had been during the second half of the eighteenth century.[75] Were tailors impoverished as a group or did some journeymen marry when they would previously have remained unmarried? We cannot answer this question. Differences were even more striking among wigmakers, 22 per cent of whom had more than 1000 *livres* in property while a large majority had nothing but what they earned from their labour.[76] Some were prosperous, owning several houses, living in a relatively opulent manner and boasting silverware and even books (two wigmakers had at the time of their death a small collection of books, a fact that was never the case among tailors). It is true that wigmakers had on the whole a relatively higher level of education. Ninety per cent of them could sign their marriage settlement in 1793-97, compared to 60 per cent of tailors. Social endogamy was the rule even for the richest: marriages were seldom contracted outside the crafts world,[77] although a few seamstresses married merchants or ships' captains as a sign of social promotion. A relatively high percentage of tailors and wigmakers married a working woman,[78] often one from Bordeaux.[79] Here is evidence of the integration of these outsiders, in a proportion as high as 90 per cent for both tailors and wigmakers. The age of marriage continued to be relatively late for tailors and wigmakers, native and immigrant alike. Among tailors, natives married at 27 years, 1 month and migrants at 31 years, 11 months. Among wigmakers the proportion was respectively 28 years, 2 months and 31 years, 1 month. Few of them practised as hairdressers[80] while wigmakers were still widespread both in name and in functions.

Here was a world whose hierarchies had been little affected by the Revolution. Former guild masters remained the richest wigmakers and their opportunities of social promotion were not much greater, if better acknowledged: social promotion now rested on ability. The new ideology of individualism, economic freedom and equality had surreptitiously advanced until guilds became an empty shell. The role of the Revolution was to formalize the transition in law, and to accelerate it. It had been possible for quite a long time to exercise a trade without actually belonging to the corporation, and many journeymen, marginalized within existing corporations, had to a greater or lesser extent constituted rival *compagnonnages*.[81] Such a cleavage was even clearer in a town like Bordeaux where the guilds played no more than a minority part in production and trade. The corporate system no longer represented accurately the reality of stratification within society.[82] Social conditions varied greatly not only according to trade and between master and journeymen, but also between masters of the same trade who were often competitors more than they were fellow members. The impact of the French Revolution on urban trades was therefore not negligible, but the changes were probably less radical than might have been expected, taking their place within a long process of change rather than constituting a sudden upheaval.[83]

Notes

1. Late eighteenth-century critics of the corporate system came to use this word to describe them. It replaced older words such as *communautés*, *métiers*, or *jurandes*. On the guild system, see E. Cornaert, *Les corporations en France avant 1789*, Paris: Les Editions ouvrièves (2nd edn, 1968).
2. F. Furet and M. Ozouf, *Dictionnaire critique de la Révolution française*, Paris: Flammarion (1988), p. 127.
3. E. Faure, *La disgrâce de Turgot*, Paris: Gallimard (1961), p. 424ff.

4. According to P. Bécamps, members of friendly societies (*confréries mutualistes*) in Bordeaux were secretly preparing themselves to resist the oppression of their employers, who in turn felt sufficiently threatened to ask for the dissolution of these societies. See his contribution to G. Pariset (ed.), *Bordeaux au XVIIIe siècle*, Bordeaux: Fédération Historique du Sud-Ouest (1968), p. 375.
5. J.P. Poussou, *Bordeaux et le Sud-Ouest au XVIIIe siècle. Croissance économique et attraction urbaine*, Paris: Editions de l'Ecole des hautes études en sciences sociales (1983), p. 651.
6. C. Higounet (ed.), *Histoire de Bordeaux*, Toulouse: privat (1980), p. 137.
7. J. Pontet (ed.), *Histoire de Bayonne*, Toulouse: privat (1991), pp. 51-2.
8. J. Pontet-Fourmigué, *Bayonne, un destin de ville moyenne à l'époque moderne*, Biarritz (1990), p. 533.
9. B. Gallinato, *Les corporations à Bordeaux à la fin de l'Ancien régime*, Bordeaux (1992), p. 286.
10. Ibid., p. 287.
11. *Almanach du commerce, des arts et des métiers pour la ville de Bordeaux et la province*, Bordeaux: Berguet (1784).
12. This was a striking difference from Bayonne where the aldermen backed the corporation system in all these trades in spite of the protests of merchants and ship owners
13. They had more to do with commerce than with production; there were for example about thirty café owners in 1762.
14. P. Loupès, *Chapitres et chanoines de Guyenne aux XVIIe et XVIIIe siècles*, Paris: Editions de l'Ecole des hautes études en sciences sociales (1985), pp. 110-11.
15. An ever-increasing amount of people started their own trade without even a licence and, in spite of the requirement to have one, introduced in 1788, controls were ineffective.
16. *Archives municipales de Bordeaux, Anciens et nouveaux statuts*, Bordeaux, 1701.
17. Gallinato, *Les corporations à Bordeaux*, p. 307.
18. Violence was one way of settling problems in the world of the trades, especially in cases of conflict between *chambrelans* and representatives of the trades, who sometimes called upon the forces of law and order. Ibid., pp. 292-4.
19. *Archives départementales de la Gironde* (hereafter *ADG*) 12 B 376. In June 1785, the inhabitants of rue Bouhaut turned against the tailors' bailiffs who had seized clothes, and chased them away.
20. D. Roche, *La culture des apparences*, Paris: Fayard (1991), p. 549.
21. This was also the case in Bayonne. See J. Pontet, 'Morale et ordre public', *Société Sciences Lettres et Arts de Bayonne*, Bayonne (1975).
22. *ADG*, C 3869, 1767.
23. No fewer than 466 seamstresses and 202 dressmakers married in Bordeaux between 1793 and 1797. See F. Baudat, 'Tailleurs, tailleuses et couturières à Bordeaux pendant la Révolution et l'Empire', Mémoire de maîtrise, University of Bordeaux III, 1990.
24. S.L. Kaplan, 'Les "faux ouvriers" de Paris au XVIIIe siècle', in *La France d'Ancien régime. Etudes réunies en l'honneur de Pierre Goubert*, Toulouse: privat (1984). In Kaplan's view their numbers have been underestimated, at least for Paris.

25. Poussou, *Bordeaux et le Sud-Ouest*, p. 115.
26. *ADG*, 12 B 384.
27. See Pariset, *Bordeaux au XVIIIe siècle*, pp. 100-103.
28. *ADG*, 12 B 384, 1788. Kaplan has shown that many trade masters failed to observe the guild regulations concerning the hiring of journeymen: S.L. Kaplan, 'La lutte pour le contrôle du marché du travail à Paris au XVIIIe siècle', *Revue d'histoire moderne et contemporaine*, 36, 1989, pp. 361-412.
29. On the corporate system, see F. Olivier-Martin, *L'organisation corporative de la France d'Ancien régime*, Paris: Sirey (1938); Cornaert, *Les corporations*; and Gallinato, *Les corporations à Bordeaux*, pp. 35-64.
30. Only the sons of masters were formally excused the complete or partial production of a masterpiece, since they were thought to have become competent enough in their father's workshop.
31. *ADG*, C 1765, C 1796 and C 1797.
32. *ADG*, C 1753. In 1787 and 1788, some apprentice boilermakers became masters after paying 200 *livres* 'in place of masterpiece'. In some cases, tinsmiths paid up to 700 *livres*, *ADG*, C 1763.
33. *ADG*, C 1739, February 1789.
34. Gallinato, *Les corporations à Bordeaux*, p. 334.
35. J. Nobilé, 'Recherches sur les barbiers-perruquiers-baigneurs-étuvistes de Bordeaux au XVIIIe siècle', Mémoire de maîtrise, University of Bordeaux III, 1970. The *intendant* advocated the creation of 60 extra indentures because of the increase in the population.
36. For the example of the tailors, see J. Lacoste-Palasset, 'Les tailleurs bordelais dans la deuxième moitié du XVIIIe siècle', Mémoire de maîtrise, University of Bordeaux III, 1973.
37. Nobilé, 'Recherches', p. 35.
38. Baudat, 'Tailleurs, tailleuses et couturières', p. 75.
39. Many of them used the local press to advertise for more customers.
40. Most of the trades were scattered all over the town, with the possible exception of the butchers, but there were some districts with a high concentration of tailors, shoemakers, goldsmiths etc., especially in the urban core.
41. J.-C. Perrot, *Genèse d'une ville moderne. Caen au XVIIIe siècle*, Paris: Mouton (1975), p. 341.
42. Baudat, 'Tailleurs, tailleuses et couturières', pp. 49-50.
43. Butchers and goldsmiths were counted among the former; tailors and shoemakers among the latter.
44. Higounet, *Bordeaux*, p. 359.
45. J. Benzacar, *Le pain à Bordeaux au XVIIIe siècle*, Bordeaux: Gounouilhou (1905), pp. 117-18.
46. B. Buna, 'Le cadre de vie bordelais au XVIIIe siècle d'après les inventaires après décès. L'exemple de la petite robe et des tonneliers entre 1690-1720 et 1764-1788', Mémoire de maîtrise, University of Bordeaux, 1992.
47. Nobilé, 'Recherches', p. 61.
48. See Cornaert, *Les corporations* and, for Bordeaux, J. Cavignac, *Le compagnonnage dans les luttes ouvrières au XVIIIe siècle. L'exemple de Bordeaux*, Paris-Genève: Bibliothèque de l'Ecole des Chartes (1969), and S.L. Kaplan, 'Réflexions sur la police du monde du travail 1700-1715', *Revue historique*, 261, 1979, pp. 17-77, or A. Farge, *La vie fragile*.

Violence, pouvoirs et solidarités à Paris au XVIIIe siècle, Paris: Hachette (1986).
49. Gallinato, *Les corporations à Bordeaux*, pp. 223-4.
50. Baudat found only one such occurrence in the inventories made by Bordeaux notaries in 1789-99.
51. Guilds had their own employment agencies that restricted the masters' freedom of choice, but which the *compagnonnages* opposed because it was their intention to control hiring. Conflicts were frequent in Bordeaux (about twenty of them took place between 1750 and 1789) and are similar to the ones described for Paris by Kaplan, 'Réflexions sur la police du monde du travail'.
52. See the regulations recorded by the *Parlement* on 12 September 1781.
53. Gallinato, *Les corporations à Bordeaux*, p. 219.
54. K. Dessans, 'Bilan des recherches sur les barbiers, perruquiers, baigneurs, étuvistes à Bordeaux sous la Révolution et l'Empire', Mémoire de maîtrise, University of Bordeaux, 1993.
55. *Archives municipales de Bordeaux*, HH 133.
56. For instance an action was brought in the shoemakers' trade by the party of the 'young' (*jeunes*) against that of the 'old' (*anciens*), September 1787. *Archives municipales de Bordeaux*, HH 94.
57. S. Cerutti, *La ville et les métiers. Naissance d'un langage corporatif (Turin, 17e-18e siècle)*, Paris: Éditions de l'École des Hautes Études en Sciences Sociales (1990), p. 32.
58. Ibid., p. 204ff.
59. Members of the watch were not numerous enough to enforce law and order properly and judicial control was fairly weak given the lack of resources.
60. Gallinato, *Les corporations à Bordeaux*, p. 15.
61. Pariset, *Bordeaux au XVIIIe siècle*, p. 376ff.
62. *Archives municipales de Bayonne*, GR 200. Petition signed by 845 masters representing 4258 individuals.
63. *ADG*, C 1789.
64. On the Bordeaux trades and the preparation for the Estates General, see M. Lhéritier, *La révolution à Bordeaux*, Paris: Presses universitaires de France (1942), vol. 1, p. 212ff.
65. *Archives municipales de Bordeaux*, D 138. The town council had asked its counterpart in Lyon about how it had responded to this problem.
66. Information flowed especially between the guilds of Toulouse and Bordeaux: *ADG*, C 1775, C 1782 in particular.
67. Dessans, 'Bilan des recherches', p. 25. Holders of master's licences had obtained the right to claim the cost of their old licence against the new *patente* tax: *Archives municipales de Bordeaux*, D 139.
68. Gallinato, *Les corporations à Bordeaux*, p. 346.
69. G. Ducaunnes-Duval, *Inventaire sommaire des archives municipales de Bordeaux: période révolutionnaire*, 4 vols, Bordeaux (1896-1929). For bakers, vol. 3, pp. 290, 416, 421, nailmakers, vol. 4, p. 109 and tailors, vol. 4, p. 26.
70. Dessans, 'Bilan des recherches', p. 70: 'most of them could only find new resources through marriage'.
71. Ibid., p. 50.
72. Baudat, 'Tailleurs, tailleuses et couturières', pp. 100-101. However, there were very few marriages in the period 1793-97 between a former

journeyman and the daughter of a master or former master.
73. Roche, *La culture des apparences*. He shows that the number of wigmakers fell from 960 in 1786 to 64 in 1791, before climbing to 2500 in 1807.
74. Baudat, 'Tailleurs, tailleuses et couturières', pp. 106-8.
75. J. Lacoste-Palasset, 'Les tailleurs bordelais dans la deuxième moitié du XVIIIe siècle', Mémoire de maîtrise, University of Bordeaux, 1973.
76. Dessans, 'Bilan des recherches', pp. 68-9.
77. There were only two tailors' marriages outside the craft world: one to the daughter of a bailiff; the other to the daughter of a former employee on the king's estates.
78. Forty-two per cent of the tailors and about the same among wigmakers.
79. This was the case for 64 per cent of immigrant tailors.
80. Eleven hairdressers married in Bordeaux between 1793, in comparison with 90 wigmakers, but this included divorcees marrying for the second time; four hairdressers and three wigmakers got divorced, which was a fairly high percentage.
81. On these issues see Cerutti, *La ville et les métiers*, pp. 200-201.
82. Which in Kaplan's view they did, as cited by Cerutti; ibid., p. 16.
83. On the question of the language of labour see M. Sonenscher, 'Les sans-culottes de l'an II: repenser le langage du travail dans la France révolutionnaire', *Annales ESC*, 1985, pp. 1087-108.

CHAPTER SEVEN

Craftsmen in the political and symbolic order: the case of eighteenth-century Malmö[1]

Lars Edgren

This essay is about craftsmen in Swedish towns in the eighteenth century. It is not, however, concerned with their economic position, but instead seeks to place them in the political and symbolic hierarchies of the towns. The political influence of master craftsmen is assessed, and an attempt is made to investigate how their position in urban society could be shaped and reflected in different symbolic ways. This is an area towards which much academic interest has recently been directed. Symbolic and ritual aspects of urban life have been studied for both the medieval and early modern period in Europe, with processions receiving particular attention for their meanings and functions.[2]

Studies of this type are valuable to historians for several reasons. On one level they may reveal information on contemporary understandings of the social order, hierarchy and status in a given society. This can help historians to move beyond the theoretical categories imposed on the past from the present. Thus, it contributes to the recent trend away from interpretations of the past rooted in economic and social categories that are somehow perceived as basic to understanding past societies. This use of ceremony requires that it be regarded as a text which interprets the social order.[3] The study of ceremony can, however, serve other purposes. It can be used as a tool to explain the creation of order and consensus in past communities. Ceremony is thus often understood in terms of its function as an integrative factor in the social order of a community.[4] Both these approaches move away from an understanding of ceremonies as mere reflections of some allegedly more fundamental reality, usually an economic one. Yet while they thus bring an independent value to the study of ceremony, they do not sufficiently highlight the way ceremonies did not merely interpret but in fact constructed the social order. Ceremonies are also about conflict and power; they are used to create boundaries between people, define who is included and who is excluded from different social groups, and thus to create identities.[5]

The purpose of this essay is to discuss the way politics and ceremonies positioned craftsmen in the social hierarchies of Swedish towns in the eighteenth century. Since the political aspects of Swedish towns have been extensively studied in previous research, this aspect of the study provides little difficulty. Symbolic representation in towns is, on the other hand, largely unexplored territory in Sweden, which means that the subject is best approached through a local study. One of the larger Swedish towns, the town of Malmö in southern Sweden, has been chosen. The choice is somewhat arbitrary, and Malmö should not be understood as in some way representative. Rather, the local study should be regarded as a way of opening up these wider issues. Forms of local government in Malmö are therefore discussed as one particular instance of a general pattern. In the section on symbolic order, we shall seek public manifestations of urban society, which can be used to reveal how social status and hierarchies were perceived, presented and shaped. The focus is on *public* events, on what would be readily visible to all inhabitants of the town.

Towns in Sweden in the eighteenth century

Sweden was a sparsely populated and predominantly agrarian country in the eighteenth century. In most of the territory, towns were few and far apart. Some one hundred towns existed, but two-thirds of these had fewer than one thousand inhabitants. Only Stockholm with its 60 000 inhabitants was a truly sizeable city.[6] Most towns were small local centres with economies based on local trade and some craft production, and agriculture was an important source of income for many burghers (*borgare*). The base for the towns' privileges was the exclusive right to pursue commerce and craft production. In practice, however, there were numerous exceptions to this principle, particularly concerning the craft monopoly. If an individual wished to practise as a merchant or a craftsman, he had to earn the rights as a burgher, and only burghers had full political rights. Even small towns could play an important role in the local economy, but in Sweden towns were seldom thriving commercial and industrial places, with the exception of towns such as Stockholm, Gothenburg and Norrköping. The towns were not only important economically, however, for they also fulfilled crucial political and military functions. They constituted the centres for county government, bishops, schools, universities, and military garrisons, while some of them served as fortresses.

Not everyone residing in a town was a burgher. Within the burgher group, the merchants and craftsmen (*handlande* and *hantverkare*) were numerically dominant, although other people with a trade could become burghers, that is fishermen (*fiskare*), shippers (*skeppare*), 'urban farmers' (often called only *borgare*), and so on. The burghers' share of the total municipal population varied considerably, but was generally larger in the very small towns and smaller in the larger towns. In the latter, approximately a third of all households belonged to burghers, while in the former, the proportion was often more than half.[7] The larger cities would thus be characterized by a more differentiated social structure, with a considerable number of higher and lower public officials (*ämbetsmän*). Employed workers often resided in the households of their employers. Significant groups of wage-workers with households of their own were only to be found in the larger cities.[8]

Craftsmen constituted the largest single group among the burghers. At the end of the eighteenth century, master craftsmen and their workers made up around 10 per cent of the urban population, a proportion which was considerably lower than comparable figures for Prussia. This underlines the relative importance in Swedish towns of activities other than production.[9] Craftsmen worked in very small workshops, apparently mostly for a local market in the towns and surrounding rural areas. The crafts were regulated by a national guild code established in 1720. This legislation superseded all previous local guild statutes and sought to regulate strictly all guild activities that were deemed detrimental to the public interest. It created a comprehensive system which in theory encompassed all Swedish urban craftsmen. One main function of the code was to ensure that all competent journeymen were accepted as masters without the traditional harassment by the guilds. The architects of the guild code also wished to prevent masters from using the guilds as economic cartels, particularly in making agreements on prices and wages. Although the law was not followed in all respects, it did seriously restrict the autonomy of Swedish guilds.[10]

Craftsmen in the urban political order

Formal political power in Swedish towns was divided between the burghers and a magistracy (*magistrat*). The latter had developed from the medieval council and was in principle elected by the burghers. It served as a court of justice but was also responsible for the administration of the town. The magistracy served both as representative of state authority and of the town's local self-government. The intermingling of state and local authority[11] and of judicial, political and administrative functions is typical of Swedish town government.

The magistracy was headed by a mayor (*borgmästare*) - or in some cases two or even more - and consisted of a relatively small number of councillors (*rådmän*). During the seventeenth century the appointment of mayors had been a subject of conflict between the royal power and the burghers. The state demanded the right of appointment in order to enhance its control over the towns. A compromise was reached after the death of Charles XII in 1718 and the fall of Carolingian absolutism. This development can be understood as marking the bureaucratization of the mayors. It was now required that the mayor have a juridical education (formally as of 1749), but the burghers maintained the right to elect the mayors from among candidates with these qualifications. However, they were to draft a list of three candidates from whom the king would select one. With this system, the towns could not be forced to acquiesce in the appointment of a totally unacceptable mayor, but the formal requirements meant that the state could uphold certain standards for them.[12]

The burghers maintained the right to elect their councillors, but a certain bureaucratization can be detected in this area as well. It became common to reserve seats for so-called 'literate' councillors, councillors with some juridical education or practical experience. These councillors were recruited from outside the burgher group, while the 'illiterate' councillors were themselves burghers. While the power of the burghers was originally exercised through a town meeting (*allmän rådstuga*), it became common in the seventeenth and eighteenth centuries to create an assembly where a few burghers represented all the others. This group was usually called the elders (*stadens äldste*) and often comprised 24 burghers.

The influence of different groups among the burghers can be seen in their access to positions in the magistracy, in the composition of the elders, and in the election of representatives in the Burghers' Estate of the Swedish Parliament (*borgarståndet i riksdagen*). The 'literate' councillors did not always have a formal education, but had often served as officials in the town administration or in the civil service. In the larger towns they formed a considerable portion of the councillors. The 'illiterate' councillors were recruited from among the burghers. The merchants were clearly dominant in this context, particularly in the larger towns. In the commercial city of Gothenburg, merchants were the only ones allowed as 'illiterate' councillors. The councillors in Stockholm had a distinctly bureaucratic background. Only about one quarter of the councillors in the capital during the period 1719-72 were burghers, most of them merchants.

Craftsmen were few among the councillors in larger or medium-sized towns, but more common in smaller towns. Councillors were often recruited from among the wealthier burghers, and kin relations often formed a link among council members and created close ties between merchants, other leading burghers and 'illiterate' councillors. The councils were thus dominated by the economic elite of the towns, although we can hardly speak of an urban patriciate. The system with 'literate' councillors also tied them to a state bureaucratic system, while the councils were also to some extent open to craftsmen and less wealthy burghers.[13]

While merchants generally formed only a small group among the burghers, they were well represented in the elders. It was common to reserve seats for different groups of burghers: often half the seats for merchants and half for craftsmen. This was the case, for example, in Stockholm. In these situations, other groups of burghers would be unrepresented. However, customs varied widely, and in at least one case the elders were dominated by farming burghers. It is nevertheless evident that merchants were over-represented in comparison with their actual numbers, and that craftsmen enjoyed substantial representation within the elders, far more than other groups among the burghers. The domination of merchants would be safeguarded even without proportional representation, since elections would normally be based on the taxed income of the burghers, and the merchants normally paid a very large proportion of municipal taxes.[14]

The towns had the right to elect representatives to the Burghers' Estate of the Swedish Parliament. This was a particularly valuable right in the period from 1719–72, when Parliament was the centre of political power and the monarch was relatively weak. Since only the burghers could vote, the 'literate' councillors were excluded. As has been stated, voting was generally based on taxed income. Even if the magistracy did not vote, members were eligible to be elected and it was very common for the towns to be represented by their mayor or a councillor. In the early decades of this period, the magistrates constituted some 75 per cent of the representatives, and only a small number of the councillors were also burghers. The position of the burghers improved towards the end of the period. Merchants were notably dominant among them, although the craftsmen did have at least some representation throughout the period. In Stockholm, three of the ten seats were reserved for craftsmen, and this guaranteed craft representation. In Gothenburg, one craftsman was usually elected despite the merchants' domination of the council. Other cities would occasionally send a craftsman as a representative. While in the early part of the period craftsmen made up about 5 per cent of the representatives, their position improved in the 1760s, and in 1771 they made up as much as 12 per cent of the Burghers' Estate. This indicates that craftsmen could occasionally act as leading figures in municipal political life in Sweden.[15]

The structure of local government has some important implications for Swedish urban craftsmen. The burghers maintained a strong influence as a group, both through the election of mayors and councillors and through representative assemblies.[16] As with the medieval legislation concerning towns, craftsmen and merchants formed the two acknowledged urban economic groups. Craftsmen's right of representation was thus securely recognized, while this was not necessarily the case with burghers who pursued activities which could not easily be classified into one of these two categories. Moreover, craftsmen's role in local government was based on their rights as burghers, not on their membership of guilds. It seems plausible to assume that the political process would tend to reinforce the individual craftsman's sense of being a burgher and craftsman, rather than being a member of a particular trade, a distinction which in this context played only a minor official role. Local government thus not only reflected, but in fact shaped craftsmen's positions in local hierarchies. The absence of oligarchal rule in Swedish towns also enabled craftsmen to use local government to improve their own position. There was an opportunity for them to play a leading role. This relative openness served to blur the otherwise marked distinction between merchants and craftsmen. In this way, the actions of certain individuals could help make social distinctions somewhat more diffuse.

Craftsmen were thus far from being the leading social group in Swedish town politics in the eighteenth century. The bureaucratic element of the mayors and 'literate' councillors constituted a notably influential group. Of the burghers, the merchants were clearly dominant, reflecting their greater wealth; this yielded a disproportionate number of votes. Yet Swedish towns were not dominated by merchant patriciates, and the craftsmen, though often of modest means, were through their sheer number able to hold an important position. Their position as burghers gave them actual political influence alongside merchants. Other groups could become burghers, but had far greater difficulties in establishing a comparable position of influence.

Malmö in the eighteenth century

In the eighteenth century, Malmö was a town of roughly 4000 inhabitants. At mid-century this made it the eighth largest city in the country. It was situated in the richest agricultural area of Sweden, and the grain trade was the economically most important activity in the town, which also had a relatively large manufacturing sector, particularly in textiles and tobacco. The crafts were, with the exception of glove production, geared towards a local market in the town itself and the surrounding rural areas. Gloves from Malmö were well known and were sold on both a national and an international market. Malmö was also important for other reasons. It was the seat of the county government and thus an important administrative centre. It was also one of the main defensive towns of southern Sweden, and was surrounded by extensive fortifications. Soldiers from the garrison played a significant role in the town.

The burghers as a whole made up less than 10 per cent of the total population. Amongst these, the number of merchants remained stable, at about 40 persons throughout the century. The master craftsmen declined in number from just above 200 in the 1720s to roughly 150 at the end of the century, constituting about half of the burgher group. A numerically significant element among the burghers were those who worked in agriculture on municipal land situated outside the fortifications. There were around 100 of them. Finally, there was a smaller group of burghers in other occupations, most notably publicans (*krögare*) and innkeepers (*gästgivare*). Besides the burghers, there was a relatively large number of officials (*ämbetsmän*), clergy, teachers and professionals. The officials belonged to the magistracy and the town government, the county government, and the customs services. Compared to the 367 burgher households, this group made up 75 households in 1726. Another 89 households comprised labourers, a few married journeymen, and others at the poor end of the social spectrum. Finally, there were the soldiers, almost 1000 in 1726, most of whom were quartered in the households of the burghers.[17] There is no doubt that the merchants, though small in number, were economically dominant amongst the burghers. Their share of local taxes was considerable. A proposal from a leading merchant in 1783 that half the town taxes should be paid by the merchants was opposed by the other burghers, indicating that the merchants' share was usually even larger. Only a few of the wealthiest craftsmen reached the level of the average merchant.[18]

Local government in Malmö in the eighteenth century

The structure of local government in Malmö provides one possible version of the national pattern previously discussed. The magistracy in Malmö had an unusually 'literate' character. In 1720 no councillor was also a burgher, and most of the councillors were appointed by the provincial governor rather than elected. After this date, when the burghers could again elect councillors, they tended to prefer people with legal education or experience. A leading merchant did, however, serve as one of the two mayors in mid-century. Towards the end of the century it became somewhat more common for burghers to be elected as councillors.[19]

The representation of the burghers was made up of an Elders' assembly of 24 burghers. Half the seats were reserved for the merchants. The craftsmen held eight seats while the remaining four were reserved for the brewers, who in Malmö were considered a separate corporate body among the burghers. The chairman of the elders was usually a merchant, although one craftsman acceded to this position during the century.[20] The organization of local government in Malmö was quite complicated. Apart from the elders, the burghers were also divided into so-called societies (*societeter*), one for the merchants, one for brewers, and one for craftsmen. This made for a very complex decision-making process, allowing the magistracy to play off the societies against the elders and vice versa. As a result the influence of the elders declined.[21]

Other members of the burghers were excluded from representation until late in the century, when first the publicans and later the innkeepers were granted representation among the brewers. The large group of agricultural burghers remained without direct political influence.[22] Those outside the burghers were in principle excluded from influence, but the fact that certain taxes were based on house ownership paved the way for other groups. From the 1740s a fourth society of non-burghers played a part in local government, namely that of the higher civil servants in the town (*cleresi- och civilsocieteten*). Finally, house owners who were not considered socially fit for this society were sometimes consulted by the magistracy.[23]

Malmö's representatives in the Burghers' Estate were usually members of the magistracy, and if not, a merchant was the most likely choice. However, craftsmen were not completely excluded. A craftsman was elected twice, in both cases towards the end of the parliamentarian regime (which occurred in 1772) when, as noted above, craftsmen's representation generally improved.[24]

Although Malmö was in some ways exceptional in its administrative organization, the general conclusions about the position of craftsmen in Sweden apply to Malmö as well. The craftsmen had a secure place within the structure of political power. Even if they were clearly subordinate to the merchants, they had a recognized position above other groups of burghers, and a craftsman could occasionally reach the highest levels of local influence.

The symbolic order of Malmö

In international research considerable attention has been devoted to processions and similar expressions of civic pride.[25] In Malmö, it appears that no such events regularly took place. The municipal account books show expenditure for public representation of the urban authority on just two occasions each year. On Christmas Eve, musicians accompanied the magistrates onto the stairs of the town hall to proclaim the Christmas peace, and in February, the magistrates' court met for service (*dompredikan*) in the main church, St Peter's. At least in the early part of the nineteenth century this event involved a procession of the magistrates from the town hall to the church one block away and a return procession after the service. During this procession the church bells tolled and musicians played on the stairs of the town hall.[26] These may not have been particularly elaborate ceremonies, but they constituted definite public manifestations of the power of the magistrates in the town. Burghers and other inhabitants seem to have been only passive spectators at these events.

One occasion that did, however, require elaborate ceremonies in the town was that of a visit of a member of the royal family. These visits were not regular events, but called for a major mobilization of local resources.[27] In 1773 King Gustavus III made a brief visit to Malmö. He was received a few kilometres outside the town by the cavalry corps of the Malmö burghers. They accompanied the king and his entourage to the city. Outside the walls the commander of the garrison and his officers received the king, and inside he was met by the county governor, civil servants, the magistrates, and some burghers. After a short speech from the mayor, the guns of the fortification fired the salutation shots. The king proceeded below a temporary triumphal arch, to the performance of vocal and instrumental music. After the king had entered the church, the burghers' cavalry and their infantry corps, deployed in the main square (*Stortorget*) one block away, fired salvos.

After the service the king proceeded to the town hall (*rådhuset*) by the main square, accompanied by the cavalry. Here he was invited by the St Canute's Guild (more attention will be given to this guild later) for dinner. During the dinner a salute was fired in the main square. Later in the day the king and members of the court were introduced as members of the guild, and finally, they all participated in a ball at the town hall. In the evening the town was illuminated and decorations with poetic mottoes were displayed in some houses. Late in the evening the king departed.[28] Other visits of members of the royal family in 1744 and 1754 appear to have followed a similar pattern.

In this context it is particularly interesting to notice how Malmö chose to present itself on this occasion. First of all, it is striking that the king is first greeted by the military commanders. It is only inside the gate that the magistrates greet him, and they are accompanied by the county government. Malmö does not here seem to be a self-governing town greeting its monarch: instead it appears that the magistrates are representatives of the king in the same way as his other military and civil servants, and are even subservient to the military.

The burghers did not present themselves as corporate bodies. For example, we do not find guilds with their insignia greeting the monarch. Rather, they appeared as a military unit, as a burghers' guard (*borgargarde*). This is suggestive. Appearing in a military guise would be one way to confirm that the burghers were subservient and loyal to the king. Yet this way of presenting the city is not unique to Malmö. In Stockholm, the burghers' guard played a prominent part at royal receptions and at the celebration of other royal festivities, and royal inspections of the guard were occasions for great festivities.[29]

Such a scenario might also reflect the very high value placed on all things military in eighteenth-century Sweden. Gunnar Artéus has argued that Swedish society in this period was militarized, not only in the sense that very large resources were spent on the military, but also in the sense that military virtues were highly valued. This is, for example, revealed in the very high position of officers in the official Swedish ranking system.[30] Thus the burghers might find it honourable to be allowed to play at being military men, sharing the virtues associated with them. The fact that the burghers' infantry were in 1773 officially guarding the fortress in the temporary absence of the garrison might have provided them with a sense of personal value to king and country.

The organization of the burghers' guard reveals some of the internal rankings among the burghers. In 1744 it was assumed that the cavalry would primarily be made up of merchants and their sons and bookkeepers. Civil servants were deemed as being on an equal footing and allowed to join the burghers in the cavalry if they wished. Merchants might also serve as officers in the infantry, of which the rank and file would be made up of craftsmen and other burghers. In that year, a councillor commanded the cavalry and the town secretary the infantry.[31] Thus the local power structure would be paralleled in the command structure of the corps, with the magistracy in the lead and the merchants dominating over craftsmen. This general division was maintained. In 1754, all the captains of the four infantry companies were merchants. Among the other officers both merchants and craftsmen appear, while the lower commanding ranks contained only craftsmen and brewers.[32] Craftsmen would not, however, be universally excluded from the cavalry, and in 1773 a baker appears as cornet.[33] Distinctions within this military formation apparently did not exclude craftsmen for the whole period, and a few attempted to enter the higher ranks.

The royal entry and its symbolism does not tell a story of urban independence and pride. Indeed, to the extent that it expresses local pride, it is the pride of demonstrating loyalty and subordination to the highest possible degree.[34]

In the description of the royal entry in 1773 the important role of the St Canute's Guild is revealed. This guild had its origin as a medieval merchant guild existing all over Danish territories. The guild in Malmö survived and would serve as the most prestigious assembly in Malmö. Already by the sixteenth century, Danish kings had accepted membership of it. After Sweden conquered Malmö in 1658, the prestige of the guild was sufficiently high to make Swedish kings and high nobles accept membership.[35] It would therefore play a crucial role in royal receptions in Malmö, and the largest room in the town hall was reserved for its use.

In a history of the guild written in 1743, it is praised for the equality it accorded to both high and low. In the sense that both nobles and non-nobles could meet in the guild the author was correct. However, the line was very clearly drawn against those who were not exalted enough to become members. Those who were accepted as members were officers, officials and members of the magistracy. Of the burghers of Malmö only merchants appear as members. But clear social distinctions were made within the guild itself. At the balls it organized, for instance, the room was divided into two sections, one with a somewhat elevated floor. Only nobles were allowed on the elevated floor.[36]

Organizing social events seems to have been the main task of the guild, but on some occasions, such as the royal entries, it would appear in a public role. Similarly, the guild would celebrate royal weddings, coronations and funerals. For example, in 1771, the coronation of Gustavus III was celebrated with a public oration at the town hall, a gun salute, and fireworks.[37]

The guild would serve to distinguish the 'highest' elements among the burghers and tie them to socially superior groups in local society. The high esteem of the St Canute's Guild would certainly impart a sense of honour to those merchants who were members, but would not foster an independent identity as burghers. The guild's major role in the ceremonial life of the town would tend to diminish any attempt at civic pride centred on the burghers. While craftsmen had limited access in many other areas, it appears that the St Canute's Guild totally excluded them and thus symbolically marked their social inferiority. The guild was a useful vehicle for the local merchant elite which could aspire upwards, while simultaneously symbolically excluding the craftsmen, the group most directly below them in the social hierarchy of the town. As we have seen, this could not be done as effectively in other arenas. The guild not only reflected the hierarchy; it was one of the most efficient instruments for creating and preserving it.

Since Malmö appears to have been lacking in processions and public displays, it is interesting to search for other ways in which the social order could be symbolically displayed. It seems fruitful to focus attention on the church, a crucial institution in society where all inhabitants of the town would meet. Open to everyone, it was clearly not open to everyone on an equal footing. Seating, for instance, was one way of making distinctions among people. In eighteenth-century Malmö one could rent a seat, and it is instructive to explore the manner in which the social order was spatially presented in church.[38]

Malmö had two main parishes. The largest was St Peter's, with a large gothic basilica from the fourteenth century as its church. The second was a German parish (*Caroli*) in which services were held in German for most of the century. This analysis pertains only to St Peter's. The most striking and readily visible aspect of seating was the gender division. All women were seated north of the central aisle, the men to the south. It was also clearly more important to rent a seat for women, but since their representation in the sources and their social standing is more difficult to analyse, only the distribution of the men can be easily discussed.[39]

In St Peter's, seats were divided into two categories at different prices. Seats in the central aisle were more expensive than those in the lateral aisle.[40] The front pew in the central aisle was reserved for the county governor, behind him were seated officers from the garrison, and only then came the mayor and the magistrates. Once more the prestige enjoyed by the military is clear. The ruling magistracy of the town would take a subordinate position in its main church.

The distribution of the remaining seats by social categories for 1728 and 1771 appears in Table 7.1. The results are in some respects extremely clear-cut. No merchant would rent a seat except in the central aisle, and with very few exceptions this was also true of officials, a group which in this respect has the same social standing as merchants. Craftsmen would normally be seated in the lateral aisle. If they moved into the central aisle they would modestly remain in the back pews, not because of the cost – all seats in the aisle had the same price – but most likely in recognition of their 'proper place'. Only occasionally would a craftsman or a brewer move into the pews in the forward half of the church. It is also noteworthy that very few outside the ranks of the burghers, except the officials, would rent a seat.

The seating provided a vivid demonstration of the hierarchy in the town which everyone could read in church each Sunday. The craftsmen once more appear as a solidly respectable group among the burghers, yet clearly subordinate to the merchants, even though a few individuals could enter into a similar standing.

Funerals were extremely important events. At no other point of the life

Table 7.1 Seating of men in St Peter's Church, Malmö, in 1728 and 1771

Social group[a]	Central aisle - front	Central aisle - back	Lateral aisle	Total
1728				
Officials[b]	10	15	3	28
Merchants	13	18	0	31
Craftsmen	2	12	27	41
Brewers	0	2	3	5
Other burgers	0	2	4	6
Unknown	6	3	13	22
Total	31	52	50	133
1771				
Officials[b]	3	7	0	10
Merchants	11	19	0	30
Craftsmen	1	9	35	45
Brewers	2	3	4	9
Other burgers	0	2	9	11
Others	0	0	4	4
Unknown	1	0	3	4
Total	18	40	55	113

a For 1771 titles appear in the list. In 1728, often only the name appears. Identification has then been made from different registers in the City Archives of Malmö.
b This group also includes people with higher education (physicians, teachers).
Source: Handlingar angående kyrka och kyrkogård 1716-1810 (1728), Räkenskaper (1771), St Petri kyrkoarkiv (MSA).

cycle would the social standing of the individual be more clearly marked. Funerals in Malmö would be performed in a great variety of ways, and with a variety of distinctions, some at least of which can be analysed through parish account books and burial books. The ceremonies had several stages. Directly following the death, there might be a tolling of the church bells for the soul of the departed (själaringning). Different numbers of bells could be used for this. On the day of the funeral the guests would assemble at the home of the deceased. The body was carried in a procession to the church, or perhaps directly to the graveyard. The church bells might be tolled when the procession assembled as well as after the service. While some bodies never entered the church, in other cases the procession would simply move through the church, possibly accompanied by music. A funeral service could also be held in the church in accordance with more or less complicated ceremonials. Finally, the body would be buried, either within the church walls or in the surrounding churchyard.[41] However, many bodies would not even come close to the church. They were buried in a burial ground near the castle (*slottsplatsen*), and in these cases it appears likely that the processions would go directly from the home to the grave.[42] These differences undeniably provide a rich language through which to express social distinctions among the inhabitants of the town.

One important issue was who would carry the body. The common practice was that members of the same corporation and status group carried it. Guild members would follow their colleagues to the grave. In Malmö, several societies were formed with the express purpose of organizing the carrying of the deceased (*likbärarlag*), and at least five such 'companies' were active in the eighteenth century. Civil officials and merchants each had one. Even if there was a large number of guilds in Malmö, craftsmen also formed such societies. One was based on the relatively large glovemakers' guild, but also accepted members from other guilds. Another was a 'German company', based on the craftsmen in the German parish. The companies would also carry non-members for a fee. They helped to confirm the impression that the town consisted of separate corporate bodies. One reason for the congregation of craftsmen in funeral companies must have been the small size of guilds in Malmö. Shoemaking was the largest trade in Malmö in 1753, and there were only 16 masters in this trade.[43] Nevertheless, the fact that craftsmen chose to organize across guild lines indicates the extent to which they identified themselves as *craftsmen* rather than members of a particular guild or trade.[44] The funeral processions would provide ample opportunity to display the standing of the deceased. The participants were ranked according to closeness of kin, or else according to rank, making the social order visible to everyone in the streets.[45]

The importance of the St Canute's Guild would be underlined at funerals. The largest bell in St Peter's was donated by the guild in 1533, and was used free of charge at the funeral of all guild members and their children. Otherwise, a fee was charged for the use of the bells, and the tolling could either be with all bells (including the St Canute bell), or with the so-called 'Sunday bells'.[46] The church also charged fees for burying the body within the church walls or in the churchyard. Table 7.2 reveals how these distinctions were used in different social groups in Malmö in 1771. The results are once more extremely evident. Within the walls of the church, almost exclusively merchants and officials were buried. Brewers and their family members were commonly buried in the churchyard. Craftsmen were also often found here, but many of them had to be satisfied with an even less exalted burial place. Finally, the other burghers and inhabitants of the town could only rarely aspire to a

Table 7.2 Social distinctions in funeral practices in St Peter's parish, Malmö, 1771

Social group	Payment for bells		Burial place		
	Yes	No	Church	Churchyard	Other
Officials	8	4	9	2	4
Merchants	0	1	6	0	0
Brewers	3	5	0	9	1
Craftsmen	3	21	1	14	18
Other burghers	0	3	0	0	6
Others	0	39	0	6	62
Total	14	73	16	31	91

Notes: In the payment for bells, only adults (over 24) are included since no payment was made for tolling for deceased children. Age and gender do not affect burial place.
Source: Burial books and account books of St Peter's in Malmö, St Petri kyrkoarkiv (MSA).

burial place in the churchyard.[47] Within the church the tombstones would commemorate the deceased. For those attending Sunday services, they would reinforce the impact of seating in revealing who formed the centre of urban life. Craftsmen would rarely be members of this select group.

There were scarcely any fixed rules on social distinctions as far as public displays in church were concerned. In principle, it was simply a matter of how much the individual was willing to pay, but this is not the whole story. Complex rules would surround the 'proper' choice of ceremonies, and those breaking the rules would most likely be censored. To some extent these rules can be gleaned from recorded behaviour: for example, it was certainly not proper for a merchant to sit in the lateral aisle of St Peter's. Some craftsmen could therefore use this symbolic expression to claim a social position among merchants and officials. Even if little can be known about the actual impact of such claims, it appears likely that this symbolism was not only a confirmation of an already recognized position: it was also a part of the way such a position was actually defined.

Final reflections

This investigation reveals a comparative lack of communal celebrations in Malmö. Those events that took place were related to visits of members of the royal family, and did not indicate pride in independence and self-rule. Rather, the pride consisted of showing subjugation in the most lavish way to the symbolic centre of the Swedish state. This is hardly surprising, for Swedish towns had neither the independence nor the size to foster an independent civic culture. This is also apparent in the precedence taken by state officials and officers in the seating in St Peter's. At royal receptions, Malmö did not present itself as consisting of different corporate bodies. A military model was used, perhaps testifying to the importance of military values in Swedish eighteenth-century society.

The role of religious processions in expressing the social order in Catholic countries is well known. Such processions did not take place in Sweden after the Reformation. The church did, however, clearly remain crucial for symbolic representation of the order of the town. The funeral processions appear to fulfil a similar function in Swedish society as that of processions at religious festivals in Catholic countries. The difference is that funerals were more individualized events, simultaneously demonstrating both the social order and the deceased person's place within it. The tolling of bells and the site of the grave were other ways of asserting social distinctions. Seating in church was used in a similar way to demonstrate the social order.

This essay has focused particularly on what these events reveal about the position of craftsmen in a Swedish town, and about their political position in general. The results suggest that the position of craftsmen in both dimensions was quite similar. As burghers in a town, they had a recognized and secure position. They were, however, clearly subordinate to the merchants, who formed the elite among burghers. 'Below' craftsmen we find other groups of burghers and almost all other inhabitants of the town, who in a political and symbolic sense were made almost invisible. What is perhaps most striking is the apparent lack of a guild presence in public displays of the urban order. The guild would most likely appear publicly in the streets at least when there was a change of aldermen, when the property of the guild would be moved from the house of the old alderman to the house of the new one, probably in a small procession.[48] Nevertheless, the lack of symbolic public representation of the guild remains striking. Even at funerals, craftsmen in Malmö tended to join in funeral companies across trade lines. Both the political and symbolic order in Malmö tended to construct an identity as being a member of a community of craftsmen rather than being a member of a particular trade.

Distinctions among burghers were to some extent fluid. It was possible for wealthy craftsmen to rise into positions of considerable political influence, and also to be socially accepted as the virtual equals of merchants. A very limited number of craftsmen could successfully use these openings. Swedish towns were not ruled by urban patriciates, and being a burgher also granted access to political influence at the national level. The St Canute's Guild also played an important social role in Malmö. The only burghers permitted to enter the ranks of this select society were merchants, and it provided them with contacts with superior status groups, emphasizing in this way their distance from other groups of burghers.

The evidence provided in this essay can be read as confirming that the symbolic order reflected a more 'basic' economic and political 'reality'. The position of craftsmen is quite similar in all three areas. There is, however, one group to which this does not apply at all: the officials employed in the local and state bureaucracy. They were almost excluded from political influence in town government (if they were not magistrates). Socially, they were at least on an equal footing with the merchants and clearly superior to other burghers.

Returning to the initial conceptual reflections of this essay, it is evident that one way to interpret the symbolic practices discussed here is to claim that they reflect an underlying reality. Another would be to discuss their function in upholding the social structure. Yet in my mind these approaches are hardly satisfactory since they reduce the significance of symbols and leave them no independent role. I find it more fruitful to see these events as linked to the accumulation of symbolic capital as expressed by Pierre Bourdieu, a resource which could be used as a power resource or converted into other capital. If we understand the events as a text, they are a text that cannot only be read for the content it reveals, but also as a text which partly shaped the social 'reality' which it appears to describe.

The St Canute's Guild in Malmö is a particularly striking example of this. It can hardly be said to have reflected an underlying reality. It was in itself a resource to be used by the local elite in forging alliances with other social groups, and in defining this elite as excluding craftsmen and other groups of burghers. This process of inclusion and exclusion would have repercussions in the everyday life of the inhabitants of Malmö. This perspective is the most fruitful in the study of urban ceremonies and rituals.

Notes

1. I wish to express my gratitude to the editor of this volume for his comments, which have certainly helped to improve the essay. It has also benefited from the discussion at the session at the International Urban

History Conference in Strasbourg in 1994 and at Bengt Ankarloo's doctoral seminar in the Department of History, the University of Lund.
2. Edward Muir, *Civic Ritual in Renaissance Venice*, Princeton: Princeton University Press (1981); Mervyn James, 'Ritual, Drama, and Social Body in the Late Medieval English Town', *Past and Present*, 98, 1983, pp. 3-29; Michael Berlin, 'Civic Ceremony in Early Modern London', *Urban History Yearbook*, 1986, pp. 15-27; Herman Roodenburg, ' "Splendeur et magnificence". Processions et autres célébrations à Amsterdam au XVIe siècle', *Revue du Nord*, 69, 1987, pp. 515-33; Simona Cerutti, *La ville et les métiers: Naissance d'un langage corporatif (Turin, 17e-18e siècle)*, Paris: Édition de l'École des Hautes Études en Sciences Sociales (1990); Barbara Hanawalt and Kathryn Reyerson (eds), *City and Spectacle in Medieval Europe*, Minneapolis: University of Minnesota Press (1994); Robert Darnton, 'A Bourgeois Puts His World in Order: The City as a Text', in *The Great Cat Massacre and Other Episodes in French Cultural History*, New York: Vintage Books (1985), pp. 107-43.
3. The title of Darnton's essay on Montpellier is particularly suggestive of this approach; ibid.
4. See for example the essay on Corpus Christi celebrations by James, 'Ritual, Drama, and Social Body'.
5. The study of Turin by Cerutti, *La ville et les métiers*, provides a highly interesting study along these lines, where the need to put ceremonies into a broad contextual framework is particularly stressed.
6. Ann-Marie Fällström and Ilkka Mäntylä, 'Stadsadministrationen i Sverige-Finland under frihetstiden', in *Stadsadministration i Norden på 1700-talet*, Oslo: Universitetsforlaget (1982), p. 178f.
7. Ibid., p. 250f; Gunnar Carlson, *Enköping under frihetstiden*, Uppsala: Almqvist & Wiksell International (1977), p. 14f.
8. Kekke Stadin, *Småstäder, småborgare och stora samhällsförändringar*, Uppsala: Almqvist & Wiksell International (1979), p. 47; Eli F. Heckscher, *Sveriges ekonomiska historia från Gustav Vasa*, II:1, Stockholm: Albert Bonniers Förlag (1949), p. 91; Thomas Magnusson, *Proletär i uniform*, Göteborg: Historiska institutionen (1987), p. 108f.
9. Carl-Johan Gadd, *Självhushåll eller arbetsdelning?*, Göteborg: Ekonomiskhistoriska institutionen (1991), p. 291f.
10. For Swedish eighteenth-century guilds, see Ernst Söderlund, *Hantverkarna* 2, Stockholm: Tidens Förlag (1949).
11. Nils Herlitz, *Svensk stadsförvaltning på 1830-talet*, Stockholm: P.A. Norstedt (1924), p. 70.
12. Herlitz, *Svensk stadsförvaltning*, p. 330f; Fällström and Mäntylä, 'Stadsadministrationen', pp. 197-200.
13. Ibid., pp. 216-21; Birgitta Ericsson, 'Stockholms administration vid 1700-talets mitt', in *Stadsadministration i Norden på 1700-talet*, Oslo: Universitetsforlaget (1982), p. 329f.
14. Fällström and Mäntylä, 'Stadsadministrationen', pp. 224-8; Ericsson, 'Stockholms administration', p. 318.
15. J.E. Nilsson, 'Borgarståndet', *Sveriges riksdag*, 5, Stockholm (1934), p. 258; Fällström and Mäntylä, 'Stadsadministrationen', pp. 228-37; Ericsson, 'Stockholms administration', pp. 333-5.
16. Apparently, Swedish burghers maintained a relatively substantial role in town government. Christopher Friedrichs stresses the limited influence of

ordinary citizens in formal politics in early modern European towns in *The Early Modern City 1450-1750*, London: Longman (1995), pp. 45-50.
17. Olle Helander, 'Stadens historia 1719-1820', *Malmö stads historia 2*, Malmö: Allhems Förlag (1977), pp. 366-81.
18. Ibid., pp. 455, 516f. Published data only allow indirect inferences.
19. Ibid., pp. 481-4.
20. Ibid., p. 499f.
21. Ibid., pp. 502-5.
22. Ibid., p. 503.
23. Ibid., p. 503f.
24. Ibid., pp. 506-12.
25. Friedrichs, *The Early Modern City*, pp. 199-200.
26. These costs appear regularly in the town's account books; Stadskasseräkenskaper Huvudserien, Borgerskapets arkiv (MSA); Ludvig B. Falkman, *Minnen från Malmö*, Malmö: Malmö Fornminnesförening (1986), p. 219f. (Falkman's book is a mixture of a history and a memoir of his childhood in the early nineteenth century).
27. The royal entry into a town was an important event in all monarchies. For literature on France, see Ralph E. Giesey, 'Models of Rulership in French Royal Ceremonial', in S. Wilentz (ed.), *Rites of Power. Symbolism, Ritual, and Politics Since the Middle Ages*, Philadelphia: University of Pennsylvania Press (1985), pp. 41-64; Lawrence M. Bryant, *The King and the City in the Parisian Royal Entry Ceremony*, Genève: Librairie Droz (1986); Frédéric Barbier, 'L'entrée royale de Louis XIV à Valenciennes', *Revue du Nord*, 69, 1987, pp. 553-61.
28. Protocol of St Canute's Guild, 12 September 1773, Knutsgillets arkiv (MSA).
29. Bertil Boëthius, *Magistraten och borgerskapet i Stockholm 1719-1815*, Stockholm: P.A. Norstedts & Söners Förlag (1943), pp. 244, 424-7, 493, 521-5.
30. Gunnar Artéus, *Krigsmakt och samhälle i frihetstidens Sverige*, Stockholm: Militärhistoriska Förlaget (1982).
31. Rådhusrättens dombok, 1744 (e.g. 28 February, 26 and 28 May), Magistratens arkiv (MSA).
32. Rådhusrättens dombok, 10 July 1754, Magistratens arkiv (MSA).
33. Rådhusrättens dombok, 2 August 1773, Magistratens arkiv (MSA).
34. It is revealing that the citizens of Turin strongly protested against an attempt by the duke to have them function as soldiers at a royal reception in 1619, since this would symbolically mark their subordination. Such an independent civic pride could not be registered in Swedish towns of the eighteenth century (and most likely not in Turin either). Cerutti, *La ville et les métiers*, pp. 113-15.
35. Lauritz Weibull, *Sankt Knuts gille*, Malmö: Allhelms förlag (1956).
36. Wilh. Flensburg, *Kort Berättelse om Det så kallade S:te Knuts Gilldet*, Lund 1743; Falkman, op.cit., pp. 207-11; Helge Andersson, 'När ståndsskrankorna föll i Knutssalen', *Elbogen*, 62, 1994, pp. 87-99. Information on membership is based on a perusal of the protocol, see e.g. 20 April 1751, 28 April 1762, and 13 January 1770. *Knutsgillets arkiv (MSA)*.
37. E.g. the protocol 11 April 1766; 24 January 1770; 29 May 1771. *Knutsgillets arkiv (MSA)*.

38. Robert Tittler tentatively suggests that church seating can be seen as an alternative to processions in reformed churches where the rich Catholic ceremonial life was seriously impoverished; 'Seats of Honor, Seats of Power: The Symbolism of Public Seating in the English Urban Community, c. 1560-1620', *Albion*, 24, 1992, p. 217. This parallel is also suggested by Keith Thomas, *Religion and the Decline of Magic*, Harmondsworth: Penguin (1973), p. 180.
39. Women are given only a general title indicating civil status, sometimes a general social standing, in the listings used here. Neither town nor church records provide the material for identifying them further. They might be traced through parish account books, but only with considerable difficulty and this was not possible within the limits of this study.
40. A description of the seats in St Peter's in Falkman, *Minnen från Malmö*, p. 107f.
41. Tariffs in Handlingar angående kyrka och kyrkogård 1716-1810, St Petri kyrkoarkiv (MSA); Falkman, *Minnen från Malmö*, pp. 137-45.
42. Place of burial is noted in the burial books of St Peter's parish.
43. Lars Edgren, *Lärling - gesäll - mästare*, Lund: Dialogos (1987), p. 329. There were 28 breweries, but this trade was not reckoned among the crafts in Malmö. It was organized in its own society.
44. *Historik över likbärarlaget Enigheten i Malmö 1729-1929*, Lund (1929); Sten Kalling, *Hantverkarnas i Malmö likbärarelag, f.d. tyska likbärarecompaniet 1710-1944*, Malmö (1946). That the members of the German company actually were craftsmen appears from a list of names in the protocol in 1752. Hantverkarnas i Malmö likbärarlag, Skråarkiv (MSA).
45. Falkman, *Minnen från Malmö*, p. 142. For a discussion of the cultural implications of funeral processions in seventeenth-century Stockholm, see Bengt Sandin, 'Education, Popular Culture and the Surveillance of the Population in Stockholm between 1600 and the 1840s', *Continuity and Change*, 3, 1988, pp. 357-90. On Swedish burial practices, Nils-Arvid Bringéus, *Livets högtider*, Stockholm: LT:s Förlag (1987).
46. It is not quite clear if any bell was used if no payment was made. This does not seem unlikely, but status would still be revealed by the number of bells used.
47. Numbers are small in Table 7.2. The same pattern is however revealed in an analysis of the account books between 1771 and 1774.
48. I have not systematically investigated guild records in search of such processions.

CHAPTER EIGHT

Women and the craft guilds in eighteenth-century Nantes

Elizabeth Musgrave

> We wish to repeal this institution which thrusts aside a sex whose weaknesses have given it greater needs but fewer resources, and by condemning it to poverty, promotes seduction and debauchery.[1]

In February 1776 Comptroller-General Turgot issued the Six Edicts, the fifth of which suppressed the corporations of Paris. The limited work opportunities available to women in a city dominated by craft guilds was one motive given for their abolition. Deregulation was short-lived, however. In May, Turgot was dismissed; in August the corporations of Paris were reinstated, although reformed. Women's - and men's - work in the urban economy remained subject to limitation, regulation and control.

Women's work in early modern towns has been the subject of much recent historical interest.[2] Since the publication of Clark's *Working Life of Women in the Seventeenth Century*, it has been recognized that females played an important role in the production, distribution and consumption of goods in the pre-industrial economy.[3] The two most important influences on women's participation in the early modern economy have been identified as patriarchy, which intensified in this period and subjected women to greater tutelage by males, and capitalism, which, by changing conditions of market production and distribution, altered the nature of women's labour force role.[4] Women were subject to greater legal constraints, which reduced the work opportunities available to them. One institutional manifestation of women's diminishing legal status was their increasing exclusion from the urban craft guilds between the fifteenth and eighteenth centuries.

In contrast to women's declining politico-legal status, recent studies of wealth and occupations in early modern French towns have shown that the numbers of female workers and business owners increased from the mid seventeenth century. Collins has shown that, in Brittany and Burgundy, the numbers of female heads of household rose in the seventeenth century.[5] Farr has calculated that, in Dijon, female artisans recorded in tax rolls rose by 779 per cent between 1643 and 1750.[6] Collins has explained the paradox by arguing that increased labour force participation caused the diminution of women's legal and political position: patriarchal society responded to the threat of women's greater economic independence by placing severe institutional limitations on their freedom of action and relegated their activities to low labour-status occupations.[7] Alternatively, Albistur and Armogathe have claimed that the increased labour force participation of women enhanced their status in the eighteenth century by contributing to a modification of the corporations, which reduced some of the limitations on female work.[8]

This chapter seeks to examine the relationship between female legal status and labour force participation through a study of the position of women in the craft guilds of eighteenth-century Nantes. Nantes was a major regional and international trade centre by the eighteenth century; its participation in the Atlantic trade led to rapid population growth from 25 000 in 1600 to 80 000 in 1790.[9] Its volume of trade expanded rapidly and it was France's second port before 1789. Production and distribution of many goods within Nantes was regulated by craft corporations, which used legal privileges to maintain trading monopolies, until their abolition in 1791. By 1722, there were 32 craft guilds, which employed 20 to 25 per cent of the adult male population, particularly in the textile, food-processing, leather-, metal- and wood-working trades.[10] The importance of the guilds as both legal and economic institutions makes their study a useful starting point for an assessment of the changing position of women in the economy of the eighteenth-century city.

The central argument of the chapter is that both the legal position and the economic participation of women in the craft guilds of Nantes underwent important changes in the eighteenth century. A survey of guild statutes and regulations for Nantes shows that women's association with the corporations altered over time. One cause of modification was the changing nature of women's labour force participation in the eighteenth-century town. This influenced women's public and institutional status, at least within the craft guilds. Finally, it is suggested that the changing nature of women's participation in the eighteenth-century urban economy was both a reflection and cause of wider changes in the market production and distribution of manufactured goods.

By the eighteenth century, the craft guild was predominantly a male organization. The master craftsman had a legal function as head of household and enterprise, he participated in public assemblies, deliberations and political actions of the guild, and he enjoyed economic privileges to produce and exchange goods prohibited to other artisans in the same city. Studies of craft guilds in early modern Europe have concluded that women participated in guild production, but always in a peripheral role. Historians of the Renaissance period have claimed that female guild membership had not always been so limited; it was from the late fifteenth century that women began to be excluded from corporations and relegated to a marginal position in urban economic activity. Davis argues that economic difficulties in sixteenth-century Lyon led to increased emphasis on the male head of household as artisan and to restrictions on women becoming entrepreneurs.[11] In seventeenth-century France, according to Collins, female legal status declined further as marriage laws and inheritance rights increasingly favoured male kin.[12] Truant argues that the power or autonomy of women in almost every major sphere of influence continued to be restricted until the end of the eighteenth century.[13]

The formal apparatus of the craft guild was used as a vehicle to limit female participation in a wider campaign to reduce market competition in the early modern city. One sign of this process, Collins argues, is the loss of voting rights of women in mixed-gender guilds. Between 1620 and 1650, for example, no women voted at the examinations for entry of either the bakers' or butchers' guild in Nantes.[14] By the eighteenth century, the role of women in urban craft guilds was strictly regulated. In a few cities women had corporations of their own, as at Rouen, with five female and five mixed-gender guilds, in which women could hold meetings, vote and elect officers.[15] Such guilds were associated only with certain sectors of the economy such as textiles and food processing. They were not typical of other French cities.[16] In most corporations masters' widows could continue their husbands' businesses after their deaths, while daughters and wives, under the legal powers of their male kin, were assistants to the male artisan. Other women, outside the guild structure, were prohibited from corporately regulated craft work.

Examination of the statutes and regulations of the Nantes corporations shows that there are similarities with this model but that women's relationship with some of the guilds changed during the eighteenth century. Broadly, women's direct relationship with the guilds of Nantes was of two types: patrimonial links, through relationship with male kin who were masters, and acquisitive links, whereby women purchased independent status to practise a craft within or allied to a corporation. The rights and functions of guildswomen with patrimonial links underwent a little modification over the period whereas women who purchased individual guild membership increased in some of the Nantes corporations.

Daughters of master craftsmen were accorded privileges in the statutes of 22 out of 32 of the corporations of Nantes by 1722. Breton customary law gave all children equal inheritance rights in their parents' estates. For the sons of guildsmen in Nantes, their patrimony included material possessions and a share in their father's office and status by privileged entry into the guild, if the son had received craft training. None of the corporations of Nantes granted masters' daughters privileged entry into the craft in their own right. The requirement of formal apprenticeship, public performance of a masterpiece test and the holding of a reception banquet excluded women from full participation. Yet none of the statutes prohibited daughters from working in their father's trade. Informal training and assistance in their father's workshop may have allowed daughters to acquire craft skills and to work as artisans in some capacity, although not as independent, publicly recognized guildswomen.

Patrimonial inheritance rights to status, and the need for the beneficiary of this legacy to exercise both a public role as guild member and an economic function as artisan, was resolved in Nantes by permitting masters' daughters to transmit privileged access to a corporation to their spouses. Real estate and movable goods passed into community property after marriage, managed by the husband. Guild entry was no exception: 19 out of 32 guild statutes allowed sons-in-law of masters to enter the corporation at a reduced rate, if they had trained as craftsmen. Thus, in 1737, the joiners set an entry fee of 80 *livres* for ordinary journeymen and 40 *livres* for the sons and sons-in-law of masters.[17]

The transmission of guild status to sons-in-law is one manifestation of patriarchy: female legal and economic privileges were permissable only within a household, under male authority. Collins argues for an increase in patriarchal powers during the seventeenth century and the guild statutes of Nantes could support this. Of 11 corporations founded before 1598, only two granted privileged access to sons-in-law while 9 made no provision for daughters' rights. Between 1598 and 1700, the statutes of 13 out of 17 guilds founded (62 per cent) included provision for sons-in-law and of the 4 eighteenth-century foundations, 3 (75 per cent) did so. The increasing frequency with which inheritance rights were regulated by the guilds from the late sixteenth century also reflects the growing importance of written records as evidence of rights and privileges in early modern France. Customary practices, which were already patriarchal and dated from the late middle ages, were committed to paper as corporations resorted more frequently to law courts to maintain their status and monopolies. The original statutes of the shoemakers (1480), locksmiths (1492) and goldsmiths (1579) do not mention sons-in-law; all three guilds reissued their regulations in the late seventeenth or eighteenth centuries with a clause regulating inheritance rights. Translation of custom to written record may also have modified practice, hardening application which may have been more flexibly interpreted in orally regulated communities. What is clear, however, is that daughters and sons had rights to establish households which enjoyed privileged legal and economic status through guild membership but that the public exercise of inherited status was a male prerogative.[18]

The eighteenth century also saw changes in women's perceptions of their inherited rights as masters' daughters, at least in some of the textile and food-processing trades of Nantes. Police records show that some women considered that their status as masters' daughters gave them economic rights to trade as independent artisans, though not to participate in the public life of the corporation. Guild officials formally opposed such claims, although the extent of their censure varied. In 1753, officers of the tailors' guild visited the house of Demoiselle Rasaux; she claimed that 'she was a master tailor's daughter and that she believed herself to be well-founded to work in the trade'.[19] Nonetheless, she was deemed to be working without guild credentials and her garments were seized. But other guildsmen tacitly accepted female claims. In 1757, three daughters of the stocking-frame knitter Saget successfully appealed to the *lieutenant-général* of police to be allowed to trade up to one year after their father's death, to utilize materials purchased while he was alive. In 1781, the Saget sisters again appear in the guild records, after a visit from the officers of the corporation. It was stated that, as daughters of a master, 'they had always contributed to the debts [funds] of the community' and that their merchandise satisfied guild requirements.[20] In this example, independent women were operating on the fringes of the guild, allowed to trade because of loosely accepted inheritance rights, manifested by a willingness to contribute to guild funds. This did not extend to formal recognition by the corporation nor to access to the public position of guild master.

The position of married women was clearly defined in Breton customary law. A wife was under the legal power of her husband; she could not plead at court nor make contracts without his consent and husbands controlled all property brought into a marriage, although they could not alienate it without their wives' consent.[21] Once married, wives were without statutory provision in all but one of the Nantes guilds. The wife of a guild member was allowed to work for him, in his craft and under his privilege, but did not enter the corporation in her own right. Many wives had separate working lives from their husbands, however, even when they lived in households under the supervision of a craft guild. In some corporations the wife was the trading artisan although her husband held guild membership. In seventeenth-century Rennes, several legal professionals married widows of mercers and then joined the guild so that their wives might continue to trade.[22] At Nantes aspirant masters had to be instructed in the craft, but in some trades training was minimal, while in others masters left their craft work to take up other professions. In the secondhand-goods dealers' (*fripiers*) corporation only masters and widows could hold formal guild status, whereas it is clear from records of statute infractions that a majority of the (prosecuted) unlicensed dealers were women. Husbands joined the guild to endow their wives with trading rights. In 1761, the wife of one Landereau was charged with dealing in secondhand goods without guild status; she claimed to have dealt in the trade for five years, to have sought membership several times and was to apply again, through her husband, when he returned from a visit to the countryside.[23]

The butchers' corporation recognized that wives' separate work functions could necessitate a degree of independence within the trade. Their statutes allowed masters' wives to trade individually, under the auspices of the guild, as vendors of offal, heads and feet of cattle and sheep.[24] The wives of artisans had to work to help support their households; sometimes this was jointly with their husband but not all guild masters were independent shop-owners, for some worked for other masters and others abandoned their trades. The wives of such men could not always participate in the craft. There was tacit corporate recognition of women's independent work within certain trades but no change in the legal and official status of wives in the Nantes guilds before 1790.

Widows were the only women granted full trading privileges and official guild membership by the corporations of eighteenth-century Nantes. This was in part because of their important inheritance rights: widows became heads of household and exercised the legal rights associated with that position; they were entitled to make contracts alone and to remarry without the permission of their kin.[25] Marital property was divided after the death of the spouse between the widow and other heirs, and widows normally had guardianship of all communal property while they had minor children. All but two of the 32 Nantes guilds allowed widows the right to continue their husbands' enterprises. Only the surgeons' guild, founded in 1758, prohibited widows from continuing their husbands' work but it could not deny women their economic rights: an annual pension of 100 *livres* was to be paid to compensate for loss of income.[26]

Widows were permitted to run workshops, to hire journeymen, to manufacture and to trade. They were members of the guild confraternity, with access to burial rites and to poor relief. In return, they paid guild dues and fees. Yet while they could participate in the economic privileges of the craft guild, widows never became full 'masters' of the guilds of Nantes and did not participate in its public and political life. Widows were not present at guild deliberations, nor were female officers elected; they did not take part in public processions, nor did they plead for the corporation in the law courts. In 1759, a widow and a master nailmaker were found guilty of overpaying their journeymen. While Master Cornu was summoned before the guild assembly to answer charges, the widow merely had to pay the fine.[27]

Restrictions were placed on widows' social and economic liberties. The moral probity of independent women was of concern to 12 of the 32 Nantes guilds, which allowed widows to trade only if 'they lived modestly and chastely'.[28] Nine of these corporations were metal-, leather- and woodworking trades, which also placed limits on widows' economic liberties. Widows did not have the right to take on male apprentices. Apprentice goldsmiths had to move to a new master's workshop if their former master died.[29] Only five of the Nantes corporations allowed widows to finish the training of apprentices in their workshops, if the latter consented, although 26 guild statutes did not mention widows and apprentices at all. Training probably continued where a woman maintained a trade, but no more apprentices were hired. Loats has observed for sixteenth-century Paris that men and women usually taught worker-learners of their own gender and only rarely of the opposite sex.[30]

The Nantes corporations differ in the freedom accorded to widows to trade. In 14 of the 32 guilds, widows were allowed to continue trading during their widowhood, hiring journeymen as they required. The majority of these crafts were textile producing trades: drapers, hosiers, serge-weavers, tailors, dyers and linen-weavers, for example, or small metal trades, such as pewterers and locksmiths, which required trained male skill for the production of artefacts but where women could finish and vend. These crafts were concerned with the secondary processing of raw materials, which could have been carried out in a household workshop. The most common restriction on widows' work was the statutory provision that she had to employ skilled journeymen to carry out the craft work within her enterprise. Fourteen Nantes guilds allowed widows to continue their husbands' enterprises if male journeyman labour was employed, although only two corporations, the bakers and the vinegar-makers, limited the size of a widow's workshop. The apothecaries' guild stipulated that a widow had to employ a worker, who should remain with her for three years if he wanted to enter the guild himself.[31] The widows of butchers might trade independently in offal, heads and feet, as wives were permitted to do, but to open a shop and to slaughter animals they had to employ journeyman labour.[32] Many of the craft guilds which required a male presence in the workshop were metal-, leather- and woodworking trades, which involved work with primary products, not all of which could have taken place in a simple household workshop. The public domain in which many of these crafts operated placed restrictions on the involvement of women. Removed from the domestic site for at least part of the time, production processes may have been more 'masculinized'; widows, as head of households, might supervise the trade but could not enter into craft function and culture.[33] Given the coincidence of statutes regulating the moral probity of widows and the requirement of male craft labour, the 'masculine' identity of these trades may have demanded a strict code of conduct in the workshop for both sexes and a more formal social relationship between them. Work roles and domestic functions may have been more segregated by gender than in other crafts.

The restrictions on trading and even on workshop size limited competition from female workers but detracted from the ability of a master's family to support themselves after his death. Some guilds recognized the difficulties of women without male craft skill. The provision of labour for needy masters and widows was one justification for the continued attempt to control the placement of journeymen in Nantes during the eighteenth century. In 1762, a *règlement de police* for the tailors' guild included the provision that 'masters and widows without workers will be preferred to those with journeymen' and in the case of a widow and a master needing journeymen, the former would be preferred.[34] The Nantes guilds were concerned to control the legal privileges inherited by widows, especially their right to transmit guild status to 'outsiders' beyond the corporation. By the eighteenth century, widows (and masters) were forbidden to farm out their privileges to other craftsmen, a long-standing customary practice: they were only to employ and pay workers directly.[35] Further, widows could not treat guild membership as their transmissable property: privileges to trade ended on remarriage, although most crafts allowed widows to bestow privileged entry rights to their new husbands, if the latter were trained journeymen. The saddlers granted new husbands the same entry conditions as sons of masters: a reduced entry fee and a simple masterpiece test. The serge-weavers reduced the compulsory length of service and the entry fee but still demanded a full masterpiece test. Corporate endogamy was strong in early modern towns amongst the widows of artisans.[36] If new husbands did not enter the guild, widows were forced to stop trading. In 1770, the widow of the master secondhand-goods salesman, Lionneau, was ordered to close her shop within 15 days when she remarried with a bailiff of the city.[37]

The participation of women with patrimonial 'rights' in the craft guilds of Nantes underwent only limited modification in the eighteenth century. Women could work under and inherit the economic privileges of their fathers and husbands but continued to be excluded from the public and political life of the corporation, as they had been since the middle ages. The economic role of women in all craft guilds was determined above all by the nature of the trade. Customs and practices which underlay guild regulations seem to have existed by the fifteenth century and they underwent little discernible alteration before 1790.[38] Customary practice determined work roles which in turn affected the conditions set by guilds for female participation.

The rights of women to purchase rather than to inherit guild status increased in eighteenth-century Nantes and comprised an important modification of their legal position in the city. After 1700, a number of corporations modified their membership to accommodate female artisans. The changes were limited in scope: women could not enter guilds as mistresses in the full sense of the word. In each known example, a separate section was created for women under the regulation and organizational structure of an existing corporation. The 'mistresses' were accorded a range of economic privileges but not public participation in the guild: they were associate rather than full members. Participation was also limited to specific sectors of the economy, to certain textile crafts, food-production trades and those concerned with female adornment: in sum, to sectors in which females were traditionally active. While this was a limited concession compared to towns such as Rouen and Paris, which both had important single-sex and mixed guilds, it was a novel and significant development for Nantes.[39]

In 1728, the tailors' guild created a mistresses' section. Single and married women over the age of 20 'of good repute and standing' could enter the corporation independent of their relationship with a male artisan. They had to pay an entry fee of 120 *livres*, for which they received the right to produce clothes for women and children. There was no formal reception into the guild although an oath was taken before the police court; women paid for economic privileges, not for rites of passage into full, adult membership of a public institution. Mistress tailors were not accorded the full economic liberties enjoyed by masters. They could not open a shop and were confined to bespoke work; they could not employ journeymen and thus compete with masters for skilled workers; and their privilege was personal, unlike that of masters, for it could not be transmitted to husbands or children.[40]

Similarly, the butchers' guild at Nantes was admitting women and men in a limited economic capacity by the mid-eighteenth century, without full corporate membership. In 1750, Marie Maillard requested that 'desiring to buy, to slaughter, to sell and to butcher pigs' she be received as a mistress pork butcher (*lardière*). This seems to have been a controlled subsidiary of the butchers' guild under the regulatory powers of their statutes and officers. Pork butchers were not full members of the guild and for this reason included independent women.[41] Daughters could follow their fathers' profession without limitations. In 1755,

Geneviève Domage, unmarried daughter of a master pork butcher, was also received into this 'sub' guild.[42]

Two other guilds, the dyers and the wigmakers, increasingly tolerated women workers. In 1757, the guild masters requested that 'all masters, mistresses and widows' should not hire journeymen unless the latter possessed *billets de congé* from their previous employer.[43] As Pied points out, the care with which status was determined shows that there were three types of guild member by this date. The barber-wigmakers frequently employed women in the eighteenth century: in a *règlement* of 1724, *tresseuses* (plaiters, braiders) were recognized as part of the workforce of masters.[44] In 1772, letters patent concerning 'women and girl hairdressers and curlers' permitted women to work independently in these trades, but not to make artificial hairpieces, to run hairdressing schools or to have apprentices. Women workers were to register with the guild of wigmakers and to pay an entry fee to be allowed to trade.[45] In 1773, single and married women hairdressers were allowed to take on female apprentices, who were to be registered with the wigmakers, and to pay a fee of 20 *livres*.[46]

The preamble to the statutes creating mistress tailors in 1728 explains why the corporation sought to widen its membership. The small number of male tailors producing female garments was cited; the growth of the female population of Nantes from the late seventeenth century, combined with growing regional and colonial trade, increased the demand for women's clothes. Second, the corporation could not police a rapidly expanding, illegal female labour force in the garment trades. The products of such workers threatened standards of production, undercut the prices demanded by master tailors and reduced their livelihood. The corporation had incurred huge debts by taking illicitly trading seamstresses to court, to the sum of 18 000 *livres*. The incorporation of women traders would force their compliance with guild statutes and reduce legal costs. Finally, mistresses would contribute to the debts of the guild, which grew with each new royal creation of offices.[47] To survive changes in market demand, new production structures and the fiscal demands of the state, the guild was forced to widen the scope of its membership if its privileges and position were to survive.

The increased status and participation of women in the corporations of eighteenth-century Nantes was primarily a result of economic growth. From the late seventeenth century there was increased market demand for both primary materials and manufactured goods, which grew in volume and diversity, at Nantes and throughout France. Nantes grew rapidly in the eighteenth century. The rapid expansion of the Atlantic trade after 1680, in which African slaves, together with domestic manufactures of textiles, luxury and metal goods, were exchanged in the Caribbean for plantation products of sugar, coffee and indigo, was the primary cause of prosperity. Coastal and riverine trade with European and French towns, particularly the redistribution of raw and refined colonial goods, but also wine, salt, fish and textiles, was important too. Finally, Nantes was a growing market in its own right. Demand for luxury goods and buildings grew with the expansion of the numbers and wealth of its mercantile elite; 230 merchants increased to 400 between 1725 and 1790, while their capital assets quadrupled.[48] The population of the city tripled between 1650 and 1790 and demand for clothing, food and other finished goods grew. In addition to numbers, Fairchilds argues that urban France experienced a 'consumer revolution' in the eighteenth century, whose main characteristic was the desire of the lower classes for 'populuxe' goods, cheap copies of elite items.[49]

The expansion of Nantes had important effects on the organization of manufacture and distribution within the city. A detailed study of the economy of Nantes has yet to be written, but three general trends may be observed. First, there was an increase in the distance between large-scale producers and small artisanal enterprises, Sonenscher's 'core and periphery' of the trades.[50] The most impressive manifestation of this was the creation of large-scale manufacturers, with royal privileges to produce and exchange luxury products primarily for export. By 1786 there were 22 refineries, 9 cotton and calico works producing 150 000 pieces of cloth, 5 rope works and manufacturers of glass, porcelain, cotton, woollen, silk and linen textiles.[51] The textile and rope manufacturers were in direct competition with guild crafts. While 600 looms and 2500 workers produced cotton cloth in the city and suburbs, 5000 workers were employed in the cotton and calico works; likewise there were numerous small rope workshops on the quays of Nantes.[52] The products and the techniques of organization of these proto-factories were a threat to artisanal systems of production.

Second, within the craft guilds of many large French towns there was increased polarity between large and small workshops. Diversity of workshop organization had always existed but it increased during the eighteenth century. Bossenga has shown for Lille that a small number of merchants or merchant-manufacturers came to dominate the wool-weaving and linen-thread industries of the town by controlling the marketing of products.[53] Likewise in Lyon and Orléans, merchants who were also guild members dominated the silk and hosiery industries of the respective towns by providing raw materials and marketing to other guild members.[54] At present, workshop sizes merely hint at differences between a 'core' of large and a mass of small-scale producers within the guilds; for example, in 1738 only 2.6 per cent of locksmiths employed more than four journeymen, whereas 81.5 per cent employed one or none at all.[55]

Finally, within all trades in eighteenth-century Nantes there was some fragmentation of production and distribution. Gaston Martin claimed that in large-scale production related to shipping - food provision, textiles and rope-making - artisanal production declined or disappeared.[56] This is overstated but Sonenscher has shown the importance of fragmentation of production processes and a growth in product specialization, such that craftsmen made either a small range of goods or parts of objects which were then assembled in other masters' workshops.[57] The distribution of goods changed; craftsmen made goods for contractors or larger entrepreneurs as much, if not more than, for sale in their own shops.

While all the Nantes crafts experienced some change during the eighteenth century, it is possible to distinguish between trades which were more or less affected. The crafts producing luxury goods and other services on a small scale for a local clientele were the least disturbed by new production techniques: goldsmiths, apothecaries, pewter-makers, saddlers, maintained small-scale workshops and a tightly regulated monopoly of production. The trades which experienced maximum change - domination by large-scale producers within and without the guild, fragmentation of manufacturing processes and changes in distribution techniques - were industries producing for the export market, for the expanding mass domestic market within Nantes, or for both of these. Textile production, particularly cotton, linen and wool, garment manufacturing, rope- and nailmaking, food and provisioning trades, woodworking trades involved in shipbuilding and furniture-making were the most affected.

Finally, all craft guilds witnessed an increase in unregulated 'illicit' production in the eighteenth century. The greatest competition to guild production occurred in the trades producing for the mass domestic market: garment, food-processing, furniture-making trades and the retailing of new and secondhand goods. The Nantes guilds traditionally had jurisdiction over the town and its suburbs, with the exception of royally privileged manufacturers and charitable concerns where orphans were trained, the Sanitat and the Hermitage. But there were also many free trades in Nantes which constantly infringed the rights of the guilds and expanding numbers of 'false' workers, experienced artisans established illicitly in hidden shops (*chambrelans*), itinerant and country producers. Bois estimates that 50 per cent of the artisans listed on *capitation* rolls were *chambrelans* working ouside the guilds, a figure which takes no account of married women and households too poor to pay the tax.[58] The records of the Nantes guilds leave the impression, which awaits quantification, that the guilds were constantly resorting to royal and police courts to prosecute infractions. The garment and secondhand-goods trades were overwhelmed by illicit producers by the end of the century.

Changes in the organization of production and distribution of manufactured goods within Nantes had two consequences for women's work within the craft guilds. First, the nature of women's work within artisanal enterprises remained largely unchanged, but shifts away from the household as the basis of production reduced the opportunity for women to work in some crafts. Second, an increase in female labour force participation occurred in those sectors of the economy which underwent the greatest expansion, the textile, garment and food-processing trades. De Vries argues that an 'industrious revolution' took place in the economies of pre-industrial Europe.[59] Population growth, declining real wages and increased taxation, together with greater acquisitiveness for movable goods, led more women and children to participate in market-oriented production. The result was some reduction in the scope, but an enormous expansion in the volume of women's labour force participation, which influenced their role in the craft guilds.

The participation of women in craft enterprises was dependent on two main factors: the physical location of work performed and the nature of the craft of the male head of household. Women were able to participate most fully in crafts performed in household workshops, using skills and techniques which were traditionally practised as part of the domestic economy. Females did not take part in work carried out on building sites and naval yards, for example, nor in work which necessitated long-distance or regular travel away from home. Among the textile, clothing and provisioning trades there was much joint work by husband and wife. Germain Hontier, tailor, made clothes for men while his wife made women's garments. In 1776, they also employed two women workers, one of whom made female clothes, in their third-floor workshop/apartment.[60] Likewise, women were important in trades concerned with adornment and health: widows of barber-surgeons ran practices after their husbands' death until legal limits were placed on women surgeons with the reform of the guild in 1758. The craft skill of almost all other manufacturing processes was a male preserve, created by formal apprenticeship and journeyman training.

There was a definite and often strict gendered division of labour, even in joint husband-wife businesses. Men took part in craft production and skilled work, hired and supervised male employees, sought and drew up new contracts for work and performed all other legal and public functions associated with the trade and the corporation. The main task of wives and widows was the provisioning of the workshop with raw materials, financial organization and the merchandising of finished goods. The Parisian glazier Ménétra's wife advanced him 300 *livres* to purchase a shop before their marriage in the 1760s.[61] Once married, she took her share of the profits of the enterprise, investing in *rentes* and accumulating savings, such that she was able to contribute 2400 *livres* to the dowry of their daughter when she married in 1785.[62] The proper role of women to retail is reflected in the butchers' statutes allowing women to trade independently in offal. Wives were assumed to be familiar with the family business. During a guild survey of the journeymen employed by master locksmiths in 1738, the wife of Master Rue was assumed to know the names and origins of the journeymen in her household/enterprise and to be familiar with guild regulations concerning their hiring and registration.[63]

But a wife was always her husband's assistant, providing subsidiary services for the better completion of his skilled work. Wives spun for their weaver husbands, polished metal for locksmiths, goldsmiths and pewterers, waxed shoes for shoemakers.[64] If a husband died and his widow continued the enterprise, she could hire and supervise workmen and act in a public capacity but she still could not fashion materials or perform skilled work in many trades. In Vendôme, the Parisian glazier Ménétra and another journeymen worked, on behalf of their widow employer, on the altar window of the church at Mondoubleau, to which the widow did not venture.[65] The masculine nature of the corporations is illustrated by the problems of disciplining journeymen by women, who had greater difficulty in imposing guild regulations without the sanction of full, male membership and authority. In November 1752, *Veuve* Jacquet was censured by the nailmakers' guild for paying her journeymen higher piece rates than those set by the guild; the men told guild officials that 'they would only have the rate which they had agreed themselves', suggesting difficulties in attracting and retaining workers.[66] The cultural and practical constraints of women working in 'male' crafts are also reflected in the limited numbers of enterprises run by widows in the craft guilds. Records of the Nantes corporations suggest that widows constituted about 10 per cent of all recorded practising masters, with little change over the course of the century.[67] Widows who continued with their husband's craft in the guilds were in a minority. Of 16 widows of the Nantes joiners' guild entitled to trade in 1775, only 6 did so. The woman who took over an apparently prosperous enterprise may have had to sell its assets to pay the husband's debts and to feed her family, leaving insufficient capital to continue the business.[68] The greatest difficulty was finding a cheap enough source of labour. Most craft guild enterprises were small, based on the labour of a master and his family.

When a master's free skills were removed, a journeyman's wages could outweigh the profits to be gained from his work.[69]

Eighteenth-century changes in artisanal production in Nantes affected the role of women in guild workshops. In most trades there was some fragmentation and specialization in the processes of manufacture and distribution.[70] Sonenscher has shown the increasing importance of sub-contracting in artisanal production, such that the craftsman made either a small range of goods or parts of objects, which were assembled in other masters' workshops.[71] The distribution of many goods changed; craftsmen made goods for contractors or larger entrepreneurs as much for sale by their wives in their own shops, if not more. These changes reduced the opportunities available to women to work in the craft enterprise. The familial nature of the workshop was also being eroded in some crafts. In 1781, the joiners' guild forbade masters and widows to give board and lodgings to journeymen, present and future. There was a move away from resident to non-resident labour, with subsequent changes in workshop culture and organization. The familiarity of wives with the practices of the trade's workforce declined.[72] But the rate of change differed for each trade and in most craft guilds it was a slow process. The average size of workshops remained small in Nantes, based in or adjacent to the household. The presence of the guilds ensured that family enterprises persisted.

The change in women's participation in the Nantes economy and in the craft guilds came from an increase in women's work for wages, a result of greater demand for and changes in the organization of the textile, garment, food preparation and retail trades. There was expansion in the opportunities for female work, both legally, under the supervision of the craft guilds, and illegally, as *chambrelante* and day workers. In the garment trades increased production in an urban economy dominated by guild monopoly was achieved by the use of outworkers, most of whom were women. Some were individual women working directly for customers at home, under guild auspices, such as Marguerite Meille, mistress, who employed several day workers in 1784.[73] But most textile outworkers produced illegally, outside guild structures. Vendors of textiles and garments sub-contracted all or part of their production to subsidiary workers in private establishments, who worked for piece rates. In 1762, Marie Lamoire, *chambrelante tailleuse* claimed to work as a day worker for another 'illegal' seamstress, Jeanne Goiaux.[74] This was forbidden by the tailors' guild's statutes: goods sold in Nantes had to be produced in the workshops of guild masters and widows. But economic growth was achieved by the fragmentation and specialization of production; guild masters and mistresses themselves profited by putting out garment production, beyond the legal limits of their own corporation. In 1772, the wife of Verger, a sailor, working as an illegal seamstress in her own home, opposed the seizure of her merchandise by the officials of the tailors' guild because they belonged to a mistress of the guild.[75] Maillard, living with her father in 1772, told the guild officers she had made the two dresses which they had seized, put out to her by a mistress tailor, who then finished the garments herself.[76] The guild attempted to reduce illicit production by taking independent women workers under its tutelage, but was increasingly limited in its attempts to control the workforce of the textile trades, including its own masters and mistresses.[77]

Hufton has observed that in eighteenth-century France the natural economy of the working classes was a family economy dependent upon the efforts of each individual member.[78] A preliminary survey of the corporations of Nantes shows that the work performed by individual family members, male and female, was prescribed by strict legal and cultural traditions, which changed little during the eighteenth century. Work was performed in accordance with definite gender roles.[79] In Nantes, most women concentrated their activities in 'women's work' - catering, selling food and drink, textile and fancy goods production and retailing, within and outside the craft guilds - for they were excluded from the preparation of most raw materials other than food and clothing.[80] This was not a new development: the Nantes economy had been characterized by separate spheres from at least the later middle ages. The organization of labour in the guild system accepted as a fundamental premise that while the family was the legal, social and economic basis of an enterprise, the kind of work performed by men was and should be different from that of women.[81] Female labour in the organized crafts was an important complement to male skill and a source of income from other occupations.

Economic change and demographic expansion in eighteenth-century Nantes had important effects on the role of women in the corporations and in the wider economy. The range and volume of both luxury and mass consumer goods and services offered for sale expanded. Production and distribution processes became more specialized and crafts increased in range and type.[82] In some trades these changes diminished the scope for partnership between husband and wife and reduced the role of the wife in certain crafts. Conversely, increased specialization within workshop enterprises and demand for consumer products encouraged the market to supply many goods previously made within the home. Many of these were women's work: clothing for women and children, laundry and food preparation.[83] As real wage levels declined in the later eighteenth century and market opportunities widened, women took up unrestricted areas of work and pushed against legal barriers to other activities. Women's greater participation in guild membership in eighteenth-century Nantes was first and foremost a result of their increased presence as manufacturers and traders within the city. Changes in corporative membership came in those guilds which were under the most pressure from structural change, the textiles and food-processing sectors and also personal services. To protect their trade monopolies, guilds such as the tailors had to broaden the basis of their membership.

The later eighteenth century also saw an attack on corporate privilege by the physiocrats, who were influential within the royal and regional administrations. There was some weakening of the authority of the guilds over urban economic life in Nantes after 1770. The political and legal institutions of Nantes were more open to the economic participation of non-privileged groups and the reduction of corporate trade monopolies. Truant argues that in Nantes Turgot's reforms were favourably received by important sections of the population and Martin asserted that 'the state of public opinion had sapped the principle of the corporation even before the law would authorise their demise'.[84] This was true among artisans as well. In December 1776, the police reported that four sons of master joiners in Nantes had set up as masters without paying fees, believing a rumour that the guilds were to be abolished. Even sons of masters with trading privileges may have found it difficult or irksome to trade within the confines of a regulated trade.[85] Vested interest did preclude the abolition of the corporations, however; fears for the quality of products and of worker insubordination - there were numerous bitter disputes between masters and journeymen in Nantes after 1760 - were important causes of the call to preserve the guilds in the municipal *cahiers* of 1789. But Nantes was increasingly like Lyon, where Garden concluded that a small number of masters continued to reinforce guild monopoly by recourse to the law, while at the same time guild authority was eroded because more and more people worked outside the guilds and refused to recognize their rules and authority.[86] Brown argues that there was an inverse relation between the ability of guilds to regulate economic activity and the extent of female participation in the labour force in early modern Europe.[87]

In conclusion, this analysis of women's role in the craft guilds of eighteenth-century Nantes has emphasized the importance of economic growth as a cause of change in methods of manufacture and distribution of consumer goods. Artisanal production in small workshops and household enterprises remained the basis of economic organization, and expansion was largely achieved through a modification and multiplication of traditional structures, at least in the period before 1790, but there was a rapid expansion in the volume of goods made and a fracturing of manufacturing processes. There was little cultural change in definitions of work roles, which remained strictly gendered. There was some public recognition of women's role in the economy by their greater participation, formal and informal, in certain of the craft guilds but no reform of their legal or political position. Economic growth in the pre-industrial town of Nantes was achieved by an expansion of existing roles, not a reformation of work identities, legal position, or shifts in social attitudes.

Notes

1. M. Albistur and D. Armogathe, *Histoire du feminisme français*, Paris: Des Femmes (1977), pp. 259-60.
2. N.Z. Davis, 'Women in the Crafts in Sixteenth-Century Lyon' in B.A. Hanawalt (ed.), *Women and Work in Pre-Industrial Europe*, Bloomington: Indiana University Press (1986), pp. 167-97; M.C. Howell, *Women, Production and Patriarchy in Late Medieval Cities*, Chicago: University of Chicago Press (1986); L. Roper, *The Holy Household: Women and Morals in Reformation Augsburg*, Oxford: Clarendon Press (1991); M.E. Wiesner, *Working Women in Renaissance Germany*, New Brunswick: Rutgers University Press (1986).
3. A. Clark, *Working Life of Women in the Seventeenth Century*, London: Routledge (1919).
4. For a recent summary of this model, see B. Hill, 'Women's History: a study in change, continuity or standing still?', *Women's History Review*, 2 (1), 1993, pp. 3-20.
5. J.B. Collins, 'The Economic Role of Women in Seventeenth-Century France', *French Historical Studies*, 16 (2), 1989, p. 465.
6. J.R. Farr, 'Consumers, commerce and the craftsmen of Dijon: the changing social and economic structure of a provincial capital 1450-1750', in P. Benedict (ed.), *Cities and Social Change in Early Modern France*, London: Routledge (1989), p. 145.
7. Collins, 'Economic Role of Women', pp. 467ff.
8. Albistur and Armogathe, *Histoire du feminisme français*, pp. 257-9.
9. P. Benedict, 'French Cities from the Sixteenth Century to the Revolution', in P. Benedict (ed.), *Cities and Social Change in Early Modern France*, London: Routledge (1989), p. 24.
10. Tableau Dressé en 1722 par la communauté de ville in E. Pied, *Les anciens corps d'arts et métiers de Nantes*, Nantes (1903), vol. I, p. 11; C.M. Truant, 'Independent and Insolent: Journeymen and their "Rites" in the Old Regime Workplace', in S.L. Kaplan and C.J. Koepp (eds), *Work in France. Representation, Meaning, Organisation and Practice*, Ithaca: Cornell University Press (1986), p. 133.
11. Davis, 'Women in the Crafts', pp. 167-97.
12. See 'Comments' by J.C. Collins in *Proceedings of the Annual Meeting of the Western Society for French History*, 16 (1989), pp. 423-7.
13. Truant, 'Independent and Insolent', p. 425.
14. Collins, 'The Economic Role of Women', p. 458.
15. D.M. Hafter, 'Gender Formation from a Working Class Viewpoint:

guildswomen in eighteenth-century Rouen', *Proceedings of the Annual Meeting for the Society of French History*, 16, 1989, p. 418.
16. Ibid., p. 418. See also Truant, 'Independent and Insolent', p. 425.
17. Pied, *Anciens corps*, III, p. 150.
18. Despite the inheritance rights of masters' daughters, early modern towns did not see a high rate of trade endogamy. In Rennes, liquid wealth rather than inherited status became increasingly important in the creation of enterprises during the eighteenth century. Acquisition of guild membership through marriage to masters' daughters declined in the latter part of the century; guild records of receptions into the glaziers' and the locksmiths' guilds do not record a single entry by marriage between 1760 and 1790. See E.C. Musgrave, 'Women in the Male World of Work: The Building Industries of Eighteenth-Century Brittany', *French History*, 7 (1), 1993, p. 38. Journeymen seem to have preferred to marry a wife with a dowry, to purchase their masterhoods and to establish marital bonds throughout the *artisanat* rather than to confine themselves to the social world of one guild, as the world of the trades became more diverse. See J.R. Farr, *Hands of Honor. Artisans and Their World in Dijon, 1550-1650*, Ithaca and London: Cornell University Press, pp. 133-4.
19. *Archives Municipales de Nantes (AMN)* HH 170. Tailleurs. Contraventions 1743-70.
20. AMN HH 108. Bonnetiers et Fabricants du Bas au Métier. Contraventions 1734-86.
21. Hafter, 'Gender Formation', p. 416.
22. Collins, 'Economic Role of Women', p. 427.
23. AMN HH 138. Fripiers. Contraventions 1744-78.
24. Pied, *Anciens corps*, I, p. 185.
25. J. Hardwick, 'Widowhood and Patriarchy in Seventeenth-Century France', *Journal of Social History*, 26 (1), 1992, p. 133.
26. Pied, *Anciens corps*, I, p. 122.
27. AMN HH 116. Cloutiers. Règlements 1749-89.
28. Statuts des Arquebusiers 1671, Pied, *Anciens corps*, I, p. 31.
29. Ibid., III, p. 21.
30. C.L. Loats, 'Gender and Work in Paris: The Evidence of Employment Contracts 1540-1560', *Proceedings of the Annual Meeting of the Western Society for French History*, 20, 1993, pp. 30-3.
31. Pied, *Anciens corps*, II, p. 75.
32. Ibid., I, p. 185.
33. Roper suggests that, in sixteenth-century Augsburg, it was amongst the metal trades that gender differentials and the exclusion of women went furthest. Roper, *Holy Household*, p. 46.
34. AMN HH 168. Tailleurs. Statuts et Règlements 1471-1781.
35. Arrêt de Parlement 1785. Pied, *Anciens corps*, I, p. 76.
36. One half of all widows who remarried in Dijon before 1650 did so within their first husband's craft. Farr, *Hands of Honor*, p. 142. The chances of any widow remarrying in the eighteenth century were limited, however. See J. Dupâquier, *La Population française aux XVII et XVIII siècles*, Paris: Presses universitaires de France (1979), p. 25.
37. AMN HH 138.
38. See Howell, *Women, Production and Patriarchy*, Conclusion.
39. Similar changes occurred in Lyon where, after years of conflict, the

parlement finally accorded women *tireuses* access to the silk-makers' craft in 1786. Albistur and Armogathe, *Histoire du feminisme*, p. 259.
40. Pied, *Anciens corps*, III, pp. 176-7.
41. AMN HH 70. Police de Nantes. Régistre des Maîtrises et Jurandes 1749-50.
42. AMN HH 73. Police de Nantes. Régistre des Maîtrises et Jurandes 1754-57.
43. Pied, *Anciens corps*, III, pp. 270-71.
44. Ibid., I, p. 49.
45. Ibid., I, p. 71.
46. Ibid., I, pp. 73-4.
47. Ibid., III, p. 176.
48. G. Duby (ed.), *Histoire de la France urbaine*, Paris: Seuil (1981), III, p. 369.
49. C. Fairchild, 'The Production and Marketing of Populuxe Goods in Eighteenth-Century Paris', in J. Brewer and R. Porter (eds), *Consumption and the World of Goods*, London: Routledge (1993), pp. 228-9.
50. M. Sonenscher, *Work and Wages. Natural Law, Politics and the Eighteenth-Century Trades in France*, Cambridge: Cambridge University Press (1989), p. 193.
51. P. Bois (ed.), *Histoire de Nantes*, Paris: Privat (1968), pp. 193-4.
52. Ibid., p. 144.
53. G. Bossenga, *The Politics of Privilege. Old Regime and Revolution in Lille*, Cambridge: Cambridge University Press (1991), pp. 152-3.
54. G. Bossenga, 'Protecting Merchants: Guilds and Commercial Capitalism in Eighteenth-Century France', *French Historical Studies*, 15 (1988), pp. 702-3.
55. AMN HH 165. Serruriers. Contraventions 1738-90.
56. G. Martin, *Capital et travail à Nantes au cours du XVIII Siècle*, Nantes (1932), pp. 86-8.
57. M. Sonenscher, *Work and Wages*, pp. 22-7, 31-4, 130-51.
58. Bois, *Nantes*, p. 172.
59. J. de Vries, 'Between Purchasing Power and the World of Goods: understanding the household economy in early modern Europe', in Brewer and Porter, *Consumption and the World of Goods*, p. 114.
60. AMN HH 171. Tailleurs. Contraventions 1772-91.
61. J.-L. Ménétra, *Journal of My Life* (transl. Arthur Goldhammer), New York: Columbia University Press (1986), p. 169.
62. Ibid., p. 181.
63. AMN HH 165.
64. L. Tilly and J.W. Scott, *Women, Work and Family*, New York: Holt, Rinehart and Winston (1978), p. 47.
65. Ménétra, *Journal*, p. 34.
66. AMN HH 116.
67. See Musgrave, 'Women in the Male World of Work', p. 51.
68. O.H. Hufton, 'Women and the Family Economy in Eighteenth-Century France', *French Historical Studies*, 6, (1975), p. 17.
69. Roper argues that the right to continue a workshop was a theoretical privilege rather than a provision which was of use to the majority of widows. Its primary function was as part of a marriage strategy, protecting the workshop until the widow remarried and transferred her guild rights to

a new husband. Roper, *Holy Household*, p. 52.
70. For details on this process at Nantes, see Truant, 'Independent and Insolent', pp. 133-6.
71. Sonenscher, *Work and Wages*, pp. 22-7, 31-4, 130-51.
72. Pied, *Anciens corps*, II, pp. 168-9.
73. *AMN* HH 171; HH 170.
74. *AMN* HH 170.
75. *AMN* HH 171.
76. Ibid.
77. For details on the organization of city textile and garment trades, see P. Earle, *A City Full of People*, London: Methuen (1994), pp. 141-3.
78. Hufton, 'Women and the Family Economy', p. 1.
79. Gender separated what was considered 'skilled' work from that which was considered 'unskilled'. Wiesner, *Women and Gender*, p. 85.
80. Earle, *City Full of People*, p. 147.
81. C.R. Friedrichs, *The Early Modern City 1450-1750*, London: Longmans (1995), p. 161.
82. At Dijon, the tax rolls of 1643 list seven different occupations in the commercial sector but those of 1750 list 34. Farr, 'Consumers, Commerce', p. 143.
83. Earle, *City Full of People*, p. 111.
84. Cited in Truant, 'Independent and Insolent', p. 140.
85. Ibid., p. 141.
86. M. Garden, *Lyon et les lyonnais au XVIIIème siècle*, Paris: Belles-Lettres (1970), p. 561.
87. J. Brown, 'A Woman's Place was in the Home: Women's Work in Renaissance Tuscany', in M.W. Ferguson, M. Quilligan, and N.J. Vickers (eds), *Rewriting the Renaissance: The Discourses of Sexual Difference in Early Modern Europe*, Chicago: University of Chicago Press (1986), pp. 212-13.

CHAPTER NINE

Worlds of mobility: migration patterns of Viennese artisans in the eighteenth century

Josef Ehmer

Migration and geographical mobility have become central themes of historical research in the last few decades.[1] Contrary to long-prevailing views stressing the immobility of pre-industrial societies, more recent work in social history has called attention to the extraordinarily high extent, as well as the rich diversity, of migrational movements in pre-industrial Central Europe.[2] Two main findings should be considered. First, there are certain types and patterns of migration which remained stable over the course of centuries. Second, in these migrational types we encounter a close association of spatial stability and mobility. Quite often people were on the move to maintain their traditional mode of life, and their mobility was guided by established customs and institutions. Migration was an essential and regular component of a relatively stable social and economic order.[3]

The geographic mobility of Central European artisans conforms very well to this picture. Artisans were an extremely mobile group and constituted a high proportion of the people on the move. As a rule during the early modern period, and particularly in the eighteenth century, the majority of apprentices and masters in the urban crafts and trades and at least three-quarters of the journeymen consisted of immigrants. Despite the fact that many urban guilds throughout the early modern period tried to restrict access and to limit the number of master artisans, they were able to recruit only a small proportion of their members from within their own ranks and had constantly to rely upon an influx from outside. Migration was an essential element of the social reproduction of the urban crafts and trades.[4]

Migration was also a key mechanism for the regulation of the artisanal labour market. Pre-industrial small commodity production might be characterized by a generally limited and highly fluctuating demand for products and services. A horizontal split of the artisanal labour force was a powerful tool for coping with that economic situation. Especially in the larger Central European cities the artisanal labour market consisted of three different groups. A relatively small and relatively stable group was formed by established guild masters who were permanent residents as freemen of the town, householders and married men. A second group also consisted of permanent residents who enjoyed a less favourable social and legal position: non-incorporated or even illegal artisans who switched between 'journeywork' and self-employment, ran hidden workshops or combined their trade with other sources of income, for instance as members of military guards. The third and largest group of the artisanal labour market consisted of more or less permanently mobile journeymen, unmarried young men who were on the road for a couple of years after finishing their apprenticeship and before getting a chance to establish themselves as master artisans. Travelling from one town to the next, tramping journeymen constituted a highly flexible, supra-regional workforce which provided an indispensable complement to the relatively fixed labour potential of guild masters. The Central European tramping system enabled urban crafts and trades to maintain supply-side restrictions in the labour market in the interest of an assured livelihood, and at the same time to respond flexibly to short- and long-term changes in demand.[5]

It was this economic function of the tramping system, in particular, which astonished an English eyewitness in the early nineteenth century. J.C. Symons, as a member of the parliamentary commission investigating the situation of the hand-loom weavers in the 1830s, had been sent to the continent where he encountered the custom of *Wanderschaft* in Austria. In one of his interviews, Peter Kennedy, a Vorarlberg cotton-mill owner originally from Manchester, explained to him that

> over almost every part of Germany, the trades of tailors, shoemakers, furriers &c. &c. are carried on by masters who employ journeymen on the *Wanderschaft*, as it is called; that is to say, workmen who go from town to town, stay a winter at one place, a summer at another, and receive generally, besides board and lodging, a certain sum weekly.

Symons presented an exhaustive report on this system to the commission, expressing the opinion that it was a highly useful economic arrangement: 'I am inclined to think that a great benefit results from this system, inasmuch as labour is far better apportioned to the demand; for these men often travel through several towns, till they find employment, and no accumulation of labour takes place, sinking wages below the general average.' In general, this experience reinforced his view that 'the social system is very different in Austria to what it is in England or in Scotland'.[6]

Pre-industrial small commodity production was also characterized by a high degree of specialization and division of labour. Therefore, urban artisanal labour markets were also structured by a vertical split. Strictly speaking, each occupation formed a labour market of its own, being dependent on a specifically apprenticed and qualified workforce. Skilled journeymen of one craft could only to a limited extent, and only in some crafts, be replaced by journeymen of other trades or by unskilled labourers. Artisanal production consisted of a variety of labour markets and corresponding migrational systems.[7]

Highly specialized crafts and trades which were practised by only a few artisans in large towns, such as goldsmiths, printers, bookbinders and the like, relied on a wide-ranging inter-urban migratory system: a small number of journeymen circulated between the few towns where their calling was established. In the mass trades such as shoemakers, tailors, cabinetmakers and the like, the migratory system was rather based on specific relations between urban and rural artisans. In the course of the early modern period, many Central European rural regions witnessed a growing social differentiation and a spread of rural crafts. Master artisans in villages and small market towns took on an apprentice from time to time, but very seldom did they have enough work to be in need of a journeyman. Thus, rural artisans produced a permanent surplus of apprenticed young men who entered the tramping system and circulated among towns, at least for a certain period of their life cycle. Some of them managed to establish themselves as guild masters in one of the towns they passed on the course of their tour; others settled into the twilight zone of casual and less respectable sectors of the urban labour market; others returned to their villages of origin to become self-employed masters.[8]

Artisanal migration in early modern Central Europe consisted of a variety of different migratory systems, depending on the economic structures and the qualification requirement of each trade. Nevertheless, there existed some common features. An important one was the institutional backbone of the tramping system which displayed an astounding continuity during the early modern period and to some degree well into the nineteenth century. By the sixteenth century at the latest, a specified number of *Wanderjahre* (years on tramp) had been made obligatory by many urban guilds, and an institutional framework had been established which offered a supra-regional network of social contact and material support to the tramping journeyman. Furthermore, beyond economic logic and guild regulations, migration and mobility had become a central characteristic of journeymen's culture, constituting an essential element of their identity, self-esteem and lifestyle.[9]

The aim of this essay is to put in concrete terms some of the general features of artisans' migration mentioned above by examining the mobility of Viennese artisans during the eighteenth century. Viennese guild archives offer rich sources for the study of various aspects of migration. This analysis makes use of quantitative data concerning the extent of artisans' migration as well as the geographical origins and the travel routes of apprentices, journeymen and master artisans, and it makes use of guild statutes, administrative surveys and other qualitative sources to analyse the institutions and customs which provided the social and cultural framework of migration.[10]

Vienna and its guilds in the eighteenth century

Any exploration of the geographic origins of Viennese artisans during the eighteenth century has to take into account the prominent position of the city within Central Europe and the peculiarities of the Austrian guild system. In the early seventeenth century, Vienna became the permanent residence of the Habsburg emperors and thus the capital of the Habsburg territories. The transformation of the Habsburgs' hereditary lands into an absolutist state made considerable progress, especially during the reign of Leopold I (1657-1705). Vienna became the permanent location of the court as well as of the various centralized state authorities. After the last Turkish siege in 1683, the city experienced rapid population growth: Vienna and its suburbs then had about 80 000 inhabitants; in 1710, about 113 000; at the time of the first modern census in 1754, the population had risen to 175 000 and at the end of the century it exceeded 232 000. Thus Vienna was by far the largest and the most rapidly growing urban centre in the German-speaking world and in the whole of Central Europe.[11]

Artisans formed an important part of the population. A very detailed trade census in 1736 counted a total of 10 829 artisans, who practised their trades under various legal titles such as guild masters, court artisans, licensed and even unlicensed masters.[12] In relation to the Viennese population, this would indicate about 64 master artisans (employers and self-employed) per 1000 inhabitants. Such a proportion was very common in smaller Austrian cities as well during the seventeenth and eighteenth centuries, and even the more precise Viennese trade censuses of the early nineteenth century indicate a similar ratio.[13] The number of apprentices and journeymen is known only for some occupations but not in total. The Austrian mercantilist Johann Joachim Becher calculated in 1674 an average of 2.4 journeymen per guild master.[14] Compared with available data for some of the guilds, this ratio seems to be rather high. The guild of the stocking-knitters, for instance, included in 1675 nine masters and 13 journeymen; the guild of the cabinetmakers in the 1750s comprised 133 masters and about 200 journeymen's jobs.[15] In all these cases, the number of apprentices is not known but it was undoubtedly less than the number of journeymen. As a general ratio for all crafts and trades, about one master to two apprentices and journeymen might not be unrealistic, but it would probably mark the upper borderline of the range. If one includes master artisans' family members and domestic servants in the calculation, one might guess that in the eighteenth century between 40 and 50 per cent of the Viennese population belonged, in one way or another, to the world of the artisans.[16]

In spite of their size, the formal political influence of Viennese crafts and trades was extremely limited. The municipal law of 1526 had established a local government that was completely subordinate to the sovereign and eliminated the political autonomy of the city. At the same time, it restricted the municipal administration to a small, oligarchic group. Craftsmen were expressly excluded from political power. Only 'property-owning citizens who do not carry on a craft trade' could be represented in the city council. This Viennese municipal law remained unchanged until the administrative reforms of Emperor Joseph II took effect. Following the new ordinance of 1783, the exclusion of craftsmen was no longer an established fact of law, but the limitation of political power to an oligarchic group of 100 *Bürger* (citizens or freemen of Vienna) - who were appointed by the city council which was, in turn, elected by them - remained in effect.[17]

The exclusion of the guilds and guild masters from formal political power, however, did not mean that they exercised no informal influence. Beginning with the reign of Emperor Mathias (1612-19), it became customary for the guilds to submit their ordinances to a new monarch ascending to the throne and to request the confirmation of their 'ancient freedoms'. The Habsburg emperors Leopold I (1658-1705), Josef I (1705-11), Karl VI (1711-40) as well as Maria Theresa (1740-80), in their capacity as princes of Lower Austria, as a rule confirmed the privileges of the Viennese guilds and thereby demonstrated their high regard for these venerable rights. Josef II (1780-90) was the first monarch generally to refuse to confirm the guild ordinances in the hope of eliminating the guilds.

However, all emperors also made use of these proceedings to carry out certain modifications to the guild ordinances, bringing them into conformity with their respective commercial policies. Guild ordinances of the seventeenth and eighteenth centuries provide evidence of strong continuity, as well as of a certain flexibility and capacity to adapt to changes. In the eighteenth century, new guilds were founded in an extremely wide variety of professions which had not previously been organized into such collective associations. In Vienna these included the silk-ribbon weavers (1707); the silk-weavers (1710); the producers of cured and smoked meats (1733), who had broken away from the association of butchers; the manufacturers of veils and gauze fabrics (1731); the producers of chocolate (1744); the dealers in miscellaneous wares in the suburbs (1748) and in the city (1761); the manufacturers of clock-cases (1765) and many others.[18] In 1770 there existed a total of 139 guilds in Vienna, which incorporated 4850 master artisans.[19] The average of 35 masters per guild hides an extremely wide range. Only eight guilds organized more than 100 members; many others not more than half a dozen.[20]

The attitude of the sovereigns towards the guilds' attempts at monopolization of the labour market was ambivalent. In Vienna, craftsmen in the service of the imperial court (*Hofbefreite*) had been exempted from mandatory guild membership since the second half of the sixteenth century, as were the members of the municipal guard (*Stadtguardia*) who practised crafts on a part-time basis. To these exceptions can be added the individual patents (*Schutzbefugnisse* or *Dekrete*) which had been introduced in the late seventeenth century. These enabled craftsmen, in return for the payment of a tax, to carry on commercial professions free of guild membership requirements. Moreover, there was a tendency on the part of the authorities to tolerate or, at least not to prosecute effectively the 'freelancers' (*Störer, chambrelans*) who set up shop without any official permission to conduct business. A 1725 decree by Emperor Karl VI allowed for a generous policy with respect to the dispensation of individual patents to thereby curtail the number of illegal craftsmen.[21]

The upshot of this long-term process of the legal differentiation of craftsmen and the increasing issuing of business licences beyond the purview of the guild system was that, in Vienna in the first half of the eighteenth century, only a minority of the independent craftsmen were organized in guilds. The results of the commercial census of 1736 show that only 32 per cent of the masters doing business in Vienna were *bürgerlich*, that is, guild masters and freemen of the town; 28 per cent possessed *Schutzbefugnisse* or similar patents, 27 per cent were *Störer*, 10 per cent were members of the *Stadtguardia*, while the remaining 3 per cent were *Hofbefreite*.[22] These data relativize to a great extent the picture of comprehensiveness and uniformity within the Viennese guild system. On the other hand, the eighteenth century saw a strong counter-tendency in the direction of legal standardization.

Beginning in the 1730s, and especially in the first years of the reign of Maria Theresa, state authorities oriented their policies towards a re-expansion of the guilds and the creation of a uniform and thoroughly organized class of master craftsmen and tradesmen. The *Stadtguardia* was dissolved and the *Hofbefreiungen* were abolished; the craftsmen affected by these measures had to be incorporated into the guilds under pressure from the authorities. The special patents, dispensation of which had massively increased since 1725 and had succeeded in bringing a portion of the unlicensed craftsmen into the system, were now issued only in exceptional cases. Craftsmen who already possessed *Schutzbefugnisse* were likewise incorporated into the guilds. The *Mercantil-Schema* for the year 1766 in Vienna shows that non-guild business licences were granted only under extraordinary circumstances and that now a larger proportion of craftsmen were organized in guilds.[23] During the latter half of the eighteenth century, the legal differentiation in the Viennese crafts and trades, which had been so manifestly visible in 1736, had been considerably reduced.

The main reason for this shift in governmental attitudes towards guilds can be seen in the fact that the state had to rely on guilds as the only available and functioning institutions. In the latter half of the eighteenth century, the guilds constituted important instruments in the furtherance of several areas of state economic and social policy: for instance, in collecting a newly created income tax, in policing travelling journeymen or in supporting poor artisans or widows. The more functions the guilds performed on behalf of the state, the greater the degree of control the state exerted on them. It had been an established rule since as early as the beginning of the sixteenth century that guild members were permitted to conduct their meetings only with the knowledge of the political authorities and in their presence. The trade ordinances of Leopold I, as well as the General Trade Ordinance of 1731, further specified the creation of the post of guild commissioner to be named by the city council and to be paid by the guilds. It seems that from the 1750s onwards, the guild commissioner became an established institution, participating in all meetings, having a voice in legal decisions and functioning as a co-executor of the guilds. Whereas merchants who were city council members occasionally held the post in the first half of the eighteenth century in Vienna, the guild commissioners after 1750 were exclusively administration officials.[24]

Despite their obvious lack of formal political power, Viennese guilds exercised some informal influence because they were integral parts of the guild system of the Holy Roman Empire. Even if imperial decrees from the early eighteenth century onwards strictly forbade any autonomous contacts between guilds and journeymen's associations of different states, there existed dense networks of information exchange throughout the German-speaking world and a rather homogenous system of labour practices as well as of social norms and cultural customs. That became very clear in the 1760s, when the Austrian authorities intensively discussed the question of whether the tramping of journeymen should be completely forbidden, as was the case – in the opinion of the authorities – in England and France, and whether journeymen should be allowed to marry.[25] Though there were several arguments against the tramping system and in favour of married journeymen, the Austrian authorities hesitated and finally refused to forbid tramping and to allow all journeymen to marry because that would have clearly offended the traditional customs of the empire-wide guild systems (*Reichszünfte*). Such an offence, the authorities feared, would lead to a state of affairs in which craftsmen from other regions of the *Reich* would no longer immigrate into the Habsburg lands or would even stop passing through Austria during their tramping tours. If, in Austrian artisans' workshops, single journeymen would work side by side with married ones, 'no journeymen from the Holy Roman Empire would immigrate into these places' and, conversely, no journeymen who had been apprenticed in Austria would have been accepted in the empire.[26] The only outcome after two decades of discussion was an imperial decree in 1780, which abolished the compulsory tramping (*Wanderzwang*) which had been mandated by many guild statutes for centuries, and some other measures which backed those journeymen – primarily in the textile trades – who wanted to marry or even had founded a family. In most of the crafts, however, migratory practices were influenced only to a very limited extent by high-level bureaucratic discussions or decisions.[27]

Geographic origins of Viennese artisans

An initial look at the geographical origin of Viennese craftsmen shows the enormous importance of immigration. Only a minority of the Viennese artisans was born in Vienna (see Table 9.1). Nevertheless, there quite clearly existed strong differences between apprentices, journeymen and master artisans, as well as differences between the various crafts and trades. The most comprehensive evidence is provided by a detailed survey on the immigration of master artisans to Vienna conducted by the authorities in 1742.[28] The survey shows that only 1160 out of a total of 4773 guild masters had been born in Vienna. Their number was far surpassed by the 1663 immigrants from 'abroad', the vast majority of whom came from non-Habsburg territories of the Holy Roman Empire, in particular from Bavaria, the Palatinate and Swabia. Lower Austria, the province surrounding Vienna, and the Alpine Lands which more or less comprise modern-day Austria, made only a modest contribution to the immigration of artisans to Vienna; the Czech Lands (Bohemia, Moravia and Silesia), whose inhabitants made up the majority of

Table 9.1 Geographic origins of Viennese guild masters, 1742

Birthplace	Number	Percentage
Vienna	1160	24
Lower Austria	970	20
Alpine Austria	478	10
Czech Lands	366	8
Hungary	100	2
German Lands*	1552	33
Others†	147	3
Total	4773	100

*Bavaria 324 (7%), Palatinate 323 (7%).
†36 from Habsburg territories.

Source: V. Thiel, 'Gewerbe und Industrie', in Altherthumsverein zu Wien (ed.), *Geschichte der Stadt Wien*, vol. IV, Vienna (1911), p. 430.

nineteenth-century immigrants to Vienna, remained far behind the German Lands for origin. Immigration from Hungary played a completely marginal role, despite the fact that the borders of the Hungarian part of the Habsburg monarchy lay a mere 30 kilometres east of Vienna.[29]

Obviously, greater or lesser geographical proximity was not primarily determinative in the case of the migration of artisans. We ought rather to take as our starting point the existence of migratory systems which link a group of rural regions and major cities with one another, whereby migrational patterns within the group display a greater intensity than those between the group and external regions. The area of Southern Germany and Austria seems to have constituted such a migratory system in the eighteenth century. Klaus Stopp's analysis of thousands of journeymen's travel documents strongly points to this conclusion.[30] The area from the Upper Rhine to Vienna, stretching over 700 kilometres from west to east, was in the eighteenth century a main focal point of the entire transregional migratory system of German-speaking craft-trades journeymen. The Danube was the most important transportation artery within this migratory system and the lands adjacent to the river constituted its core zone. Around this central core was a group of outlying regions such as the Central German states of Hesse and Saxony, the Czech Lands and Hungary. Of course, artisans' tramping routes took them far beyond this zone of intense migration, encompassing the entirety of German-speaking Europe and, in individual cases, even more remote areas as well. Despite the fact that Vienna was situated on the extreme eastern edge of what we might call a Southern German–Austrian migratory system, the city was an attractive destination within it due to its size and its character as the capital of the Habsburg monarchy and the *Residenz* of the emperor. A connective element within this migratory system was certainly the language. The various Alemannic, Bavarian, Central German and Austrian dialects spoken within this region could be clearly understood by migrant artisans.[31]

Of even greater significance, however, are probably the socioeconomic structures of the rural areas within the migratory system. Rural crafts and trades became extremely widespread in the eighteenth century, particularly in south-west Germany; this was the result, among other factors, of the diminishing size of peasant holdings due to partible inheritance and long traditions of proto-industrial textile production. In about 1800, Baden and Wurtemberg showed the highest proportion of rural artisans within the whole of Germany.[32] A ratio of between 50 and 70 master artisans per 1000 inhabitants, as can be found in many south-west German rural regions, was more characteristic of proportions to be found in urban centres in Central Europe. In late eighteenth-century Bavaria, as well, about three-quarters of all artisans lived in villages or small market towns.[33] On the other hand, the average size of the rural workshops was extremely small. The areas of high density of rural artisans thus continuously produced a surplus of young men who did not find employment in their home villages and filled the labour market of large cities.

In 1742 all Viennese guilds showed remarkable percentages of masters born outside the city, but the degree to which they were open to foreigners differed considerably (see Table 9.2). While in some of the guilds Viennese-born masters formed strong majorities, they were a tiny minority in others. Most of the sword-cutlers were born in the city, but almost no brewers. These differences accentuate strong interrelationships between geographic mobility and social mobility and indicate the varying attempts and abilities of the guilds to control access to their ranks. Some of the very small and some of the very rich guilds were especially successful in developing strategies to keep their numbers limited and to ensure that the offspring of guild masters received preferential treatment in filling guild openings. The large, poor and rapidly expanding crafts, on the other hand, were open in social and geographical terms.[34]

It was the 'giants' among the crafts, such as the tailors, shoemakers and cabinetmakers, in which only 10 to 15 per cent of the guild masters were born in Vienna. These were also the guilds whose efforts to control their labour market met with the least success and where competition

Table 9.2 Percentage of Viennese guild masters born in Vienna, 1742

Guild	No. of guild masters	% born in Vienna
Sword-cutlers	36	70
Bagmakers	13	61
Bookbinders	18	56
Gardeners	113	55
Goldsmiths	116	48
Butchers	32	47
Silk-weavers	29	34
Bakers	102	26
Merchants, shopkeepers	260	23
Coopers	67	22
Shoemakers	555	15
Cabinetmakers	140	14
Tailors	640	13
Weavers	31	10
Beersellers and innkeepers	361	8
Locksmiths	82	6
Brewers	70	1
All guild masters	4773	24

Note: The specific guilds listed are a selection of Viennese guilds and contained only 2665 of the 4773 guild masters in the city.
Source: V. Thiel, 'Gewerbe und Industrie', in Altherthumsverein zu Wien (ed.), *Geschichte der Stadt Wien*, vol. IV, Vienna (1911), p. 431.

between guild masters and non-incorporated masters was the strongest. In several of these trades, the number of *Störer* approached or even exceeded that of guild masters.[35] Extremely scanty information is available to us concerning these non-incorporated artisans, but it can be assumed that the proportion of immigrants among them was not below that of the guild masters. Alongside those admitted to the guilds, immigrants were obviously also absorbed into the unregulated and illegal sector of the artisanal labour market.

To be sure, the mass-scale craft trades such as the tailors, shoemakers and cabinetmakers dominated the overall statistics regarding place of origin, as is clearly shown in Table 9.2. It was precisely these crafts, which were extremely widespread throughout the countryside, which played a quantitatively important role in the migration to major urban centres. Smaller, more specialized crafts, those which produced luxury goods or were technically demanding, on the other hand, were rarely encountered even in those rural regions which were the dominant places of origin of Viennese artisans. In craft trades such as these, special migratory systems emerged, such as those linking major cities. Most of the journeymen bagmakers, for instance, came from large Saxon cities or, more generally, from capital cities of some of the minor German states.[36]

Cross-sectional statistics such as the survey of 1742 constitute, of course, only a freeze-frame image within a much longer trend and they provide no information as to the direction in which the trend was moving. Perhaps immigration played an especially large role in the reproduction and expansion of Viennese craft trades in the late seventeenth and early eighteenth centuries, as the city's population began to recover from the losses caused by the Black Death (1679/80) and the Turkish siege (1683) and its dynamic growth as an absolutist imperial *Residenz* gained momentum. The cabinetmakers' guild, for instance, shows an increase in the number of native masters during the eighteenth century: 1700-37, just 7 per cent of the guild masters were born in Vienna; 1738-48, 14 per cent; 1749-74, 25 per cent.[37]

This developmental process can be seen even more clearly among the silk-weavers, whose guild was founded in 1710 by a very small group of artisans. Six masters had laid out the proposed guild statute, submitted it to the provincial authorities and finally received their confirmation by Joseph I. In the statute these six masters are specially referred to as 'the prime initiators and *fundatores* who, in respect of the imperial liberties now granted to them, have assembled together these articles and points'.[38] In the year following its foundation, five additional masters joined the guild. Of these 11 guild masters, five belonged to a single Italian family, the Locatellis from Bergamo. Four others were of Italian origin as well: Bernardi from Venice, Cassaretto from Genoa, De Maso and Bollini from Northern Friuli. Only two masters of the founding generation bore German names and only one of them was definitely a native of Vienna.[39] During the next two decades, Italian immigrants maintained their dominance over the guild, such that, in the 1720s, the magistracy's commissioner for guilds reported that those attending guild meetings had to be well trained in the 'Roman language' (*'wallische Sprache'*) in order to understand anything.[40] In any case, over the course of the century, the Italian silk-weavers became native Viennese themselves. The first Viennese-born Locatelli became a master in 1742 and in the very same year the overall proportion of natives of Vienna in the guild had risen to 34 per cent (Table 9.2). In the second half of the century, silk-weaving experienced enormous growth, becoming Vienna's most important industry by the end of the eighteenth century and attracting artisans from all over Europe: London, Lyon, Berlin and Pest were among the birthplaces of Viennese silk-weavers during the second half of the eighteenth century. Nevertheless, in the process of institutional establishment and economic expansion, the guild increasingly drew its new members from Viennese natives. Of the 80 silk-weavers who became masters between 1751 and 1779, and whose exact birth places are known, 65 were born in Vienna.[41]

The example of the silk-weavers in the early years of their guild shows that a high proportion of immigrants was not necessarily an indicator of high social openness and that family relationships constituted an important element in the migratory systems within individual professions and/or ethnic groups. This is even more readily apparent with respect to chimney-sweeps. As in the case of the silk-weavers, this was a relatively young guild, founded in 1664 by nine masters and confirmed by Leopold I in 1670.[42] In contrast to the silk-weavers, though, the chimney-sweeps managed to keep the guild closed and to limit their number to a maximum of 18 master artisans throughout the eighteenth century. In fact, 12 to 14 families controlled the whole business and made sure that all master positions which became available were filled by members of their own ranks. Among a total of 154 successions in the trade from 1700 until 1850, a total of 135 (88 per cent) involved family members or relatives. In only 19 instances was a business sold to a non-related person, who in almost all cases came from one of the other dominant family clans.[43] Virtually all these families originated from three or four neighbouring valleys in Ticino and Grisons in Southern Switzerland and nearby Italy. Particularly noteworthy is the fact that their establishment in Vienna did not alienate them from their region of origin. Quite the contrary, for they kept up intensive relations. The Viennese chimney-sweeps recruited most of their apprentices and journeymen from the area north of Lugano. If the young fellows did not behave themselves they were sent back home. There were quite frequent marriages between families who had moved to Vienna and those who had stayed behind, as well as financial transactions such as lending and borrowing money between kin in Ticino and in Vienna. Some of the Viennese masters returned home after their retirement, especially if they had inherited land or houses.[44] The proportion of 47 per cent born in Vienna in 1742 conceals these close connections.

Geographic origins of Viennese apprentices and journeymen

All these examples show that within an enormous overall immigration into the Viennese crafts and trades, there existed a variety of quite distinct migrational patterns. An examination of the geographic origins of the apprentices adds an additional facet to this picture (Table 9.3). In contrast to the masters and journeymen, the apprentices were recruited to a great extent from Vienna itself. Lower Austria, the province surrounding the city, also constituted an important reservoir of young boys who were taken on as apprentices by Viennese guilds. The example of the bookbinders shown in Table 9.3, a trade in which more than three-quarters of the apprentices hailed from Vienna or from Lower Austria, is a thoroughly representative case. In other Central European regions, too, the migration of apprentices was a typical example of short-distance migration.[45] Nevertheless, we find a surprisingly high proportion of immigrants among Viennese apprentices as well. This applies not only to particular guilds such as the chimney-sweeps, where the trade's close-knit relationship to Ticino, Grisons and neighbouring Italy also encompassed the recruitment of apprentices. This was also true in a mass trade such as the cabinetmakers. In the first half of the eighteenth century, less than half of the apprentices were born in Vienna and the surrounding countryside, while almost a third had travelled hundreds of kilometres from Alpine regions to Vienna. This is even more surprising in light of the fact that most of the apprentices were quite

Table 9.3 Geographic origins of Viennese apprentices (%)

Region	Bagmakers 1724–85	Bookbinders 1690–1750	Cabinetmakers 1700–40	Chimney-sweeps 1770–99
Vienna	53	49	33	23
Lower Austria	–	30	15	12
Others*	47	21	52	65
Total %	100	100	100	100
Number	73	185	789	151

*Cabinetmakers: 29% of the total from Bavaria, Swabia, the Tyrol and Switzerland; chimney-sweeps: 54% of the total from Southern Switzerland (Grisons, Ticino) and Italy; bagmakers: all from large Austrian, Bohemian and German cities.

Sources: WStLA, Innungen Bücher 3/1 (Meisterbuch 1689–1773); WStLA, Innungen Bücher 42/6 (Lehrjungenaufdingbuch 1770–1864); WStLA, Innungen Bücher 54/1 (Conference paper, Annemarie Steidl); H. Zatschek, *550 Jahre jung sein. Die Geschichte eines Handwerks*, Vienna (1958), p. 130.

young. Data from several Viennese guilds in the eighteenth century shows, with a high degree of concordance, that the vast majority of apprentices were between 13 and 18 years old and had begun their training at the age of 12, 13 or 14. We do not know exactly how these young boys managed to get to Vienna, but we might assume that most of them travelled as single individuals or that they joined up with older journeymen.[46]

The beginning of their period of training meant a short phase of spatial stability for these young artisans. Most Viennese guild statutes prescribed an apprenticeship which lasted between three and five years, and up to six years in a few exceptional cases. We know, furthermore, from apprentice registrations that these prescribed durations were generally adhered to. Indeed, not all apprentices successfully completed their training. The percentage of apprenticeship drop-outs among the wood-turners from 1776 to 1782 came to approximately 68 per cent; in the case of the bookbinders during the period 1690–1750, it was 83 per cent.[47] Apprenticeship was by no means a phase of substantial stability.

The completion of the apprenticeship and advancement to the rank of journeyman, however, marked the beginning of the most mobile stage in the life of a Central European artisan. The so-called *Wanderzwang* (compulsory migration) had for centuries been a well established feature in most Central European guild statutes: only after he had spent a certain number of years tramping would a journeyman be accepted as a master. As a rule, two to four *Wanderjahre* were required.[48] As with all prescribed norms, some exceptions were granted, but in most crafts and trades the vast majority of young journeymen left the city and the familiar workshop of their master for some length of time. Over the course of the centuries, this period spent tramping had progressed beyond all normative requirements to attain the status of a custom and to become an essential component of the journeyman's identity. *Wanderschaft* certainly did not constitute a lifetime prospect but was generally regarded as a phase in one's life. The statistical analysis of the documents (*Kundschaften*) of a large number of tramping artisans in the late eighteenth and early nineteenth centuries in Central Europe has shown that around two-thirds of all migrant journeymen were between the ages of 19 and 24.[49] This does not rule out the possibility that some journeymen, even by a relatively advanced age, were unable ever to establish themselves in a permanent location and continued to tramp throughout their artisanal careers. Their numbers, however, were low.

Migration of journeymen differed considerably from that of apprentices. It was not a move from their place of origin to some specific destination city, but rather a multi-year phase of circulation in which short-term stints of employment in one city or another alternated with days or weeks spent literally on the road. The places of origin of Viennese journeymen (Table 9.4), therefore, show only one tiny frozen moment within a system of more or less permanent mobility. Only a small percentage of migrant journeymen found employment in Vienna and were entered as immigrants in the guild records which are our sources. The case of the cabinetmakers serves as an illustration. In the 1760s, the guild of the cabinetmakers included 142 master artisans and an average of 'about 200 journeymen' at work.[50] The turnover of journeymen was high. Each year in the 1760s, about 600 newly-arrived journeymen found work, three times the average number of available jobs. Many more arrived in Vienna without getting a job and thus had to continue their tramp after a couple of days spent looking for employment or visiting the tourist attractions of the city. For the eighteenth century, we do not have data to allow us to calculate the relation between those tramping journeymen who got a job in Vienna and those who did not succeed. From various other Central European

Table 9.4 Geographic origins of Viennese journeymen (%)

Region	Bookbinders 1738-49	Cabinetmakers 1738-48	Bagmakers 1815-29	Silk-weavers 1711-80
Vienna	16	6	6	57
Lower Austria	5	11	2	12
Alpine Austria	13	13	9	2
Czech Lands	12	18	11	8
Hungary	13	5	5	1
German Lands	37	45	65	11
Others	4	2	–	9
Total %	100	100	100	100
Total cases	284	1258	151	1031
Total incl. unknown origins	290	1595	153	1282

Sources: WStLA, Innungen Bücher 3/5 (Examinationsbuch 1738-1774); WStLA, Inningen Bücher 54/9; H. Zatschek, *550 Jahre jung sein. Die Geschichte eines Handwerks*, Vienna (1958), p. 134; M. Bucek, *Geschichte der Seidenfabrikanten Wiens im 18. Jahrhundert (1710-1792)*, Vienna: Verband der Wissenschaftlichen Gesellschaften Oesterreichs (1974), p. 111.

examples and from nineteenth-century Viennese evidence, one might estimate that about a quarter to a half of the tramps found employment.[51] If that ratio were true for the eighteenth-century Viennese cabinetmakers as well, then between 1200 and 2400 would have passed through the town yearly in the 1760s – that amounts to between six and 12 times the average number of jobs. The same proportion would be less impressive in absolute terms in those trades which employed one or two dozen journeymen on average. Of course, all these employment rates varied considerably depending on local and transregional business cycles.

Despite these methodological problems, the data pertaining to place of origin in Table 9.4 allow us to grasp the dimensions of migration by journeymen. The examples of the bookbinders and the cabinetmakers are to a certain extent representative for the Central European crafts and trades in general. Wherever we have quantitative data for the eighteenth century, they show that, as a rule, less than 15 per cent of all urban journeymen were born in the town in which they were registered.[52] At the same time, Table 9.4 makes it quite clear that this general rule was not equally valid for all crafts. The silk-weavers, with more than 50 per cent Viennese-born journeymen, constitute a clear exception. In the course of the century, silk-weaving had become a proto-industrial activity based on a combination of centralized manufactures and the putting-out system. Large segments of the labour force consisted of married journeymen engaged in domestic production side by side with their wives and children. A native proto-industrial proletariat partly replaced the tramping journeymen.[53]

Institutional and cultural framework of the tramping system

The tramping system of journeymen was backed by two major institutions: *Geschenk* and *Herberge*. Both concepts are equivocal. In the German language *Geschenk* means gift, donation or gratuity, but the linguistic stem of the word also includes *Schenke*, meaning inn or pub, as well as *einschenken*, to pour out, such as a glass of wine. *Herberge* was a lodging house for tramping journeymen as well as a house of call, and it was also the inn where journeymen and/or master artisans of one or more trades usually congregated.[54]

One type of source material which provides us with a glimpse into the meaning of *Geschenk* and *Herberge* are 'journeymen's ordinances', which were enacted in the early modern period by many guilds and in the eighteenth century by governmental authorities as well. Their purpose was to regulate the practices of journeymen and their interactions with masters, including wages, working hours and work discipline. The Viennese ordinance for journeymen stocking-knitters dated 10 March 1675 describes in several paragraphs a ritual that began when the travelling journeyman entered the town. As is also the case with most other ordinances from this period, many of the formulations used here are unclear and difficult to interpret, but they allow us to assess the significance of *Geschenk* and *Herberge*.[55]

Upon arrival in Vienna, the tramping journeyman stocking-knitter sought out the *Herberge* of his craft. Here he was first to obtain his 'one half-measure of wine and *Kreuzer*-worth of bread' from the innkeeper or house attendant, whereupon he was to wait until two o'clock in the afternoon.[56] This was the appointed time to 'have the *Altgesellen* [senior journeymen] sent for' and to order a second half-measure of wine. *Altgesellen*, usually two in number, were elected to serve fixed terms as representatives or spokesmen of the journeymen. Among their responsibilities was the administration of the *Gesellenlade* (journeymen's chest), a wooden cabinet in which the journeymen's cash funds and important documents such as the ordinance were deposited for safe keeping. *Altgesellen* had to organize all official gatherings of journeymen and to conduct these meetings; they represented the journeymen's interests before the masters and the governmental authorities; and usually they were responsible for labour placement. In many ordinances, we find the stipulation that consultations with the *Altgesellen* were not to take place before two o'clock in the afternoon - to keep at least a part of their working day free from the demands of their office.

As soon as the *Altgesellen* appeared in the *Herberge*, they were to exchange 'greetings and welcome' with the newly arrived journeyman, enquire as to his wishes and intentions, and order a third half-measure of wine to drink a toast to one another. Thereafter, the *Altgesellen* took leave to make their 'rounds': they went from master to master, in a strictly established order beginning with the oldest and proceeding to the youngest, advising them of the new journeyman's arrival and enquiring whether a man was needed in one of the workshops. On completion of their rounds, the *Altgesellen* returned to the *Herberge* for the next round of wine. If they had been successful in finding work for the new arrival, the *Altgesellen* escorted him that evening to his new master, at whose premises he would also be provided with room and board.

The expenses incurred by the migrant journeyman and the *Altgesellen* in the *Herberge* constituted the first portion of the *Geschenk*: the costs for the wine and bread, the so-called *Auslösung* ('ransom'), was paid for out of the 'journeymen's chest'. If the migrant journeyman found employment in Vienna, he would then receive his actual *Schenk* or *Einschenk* (welcoming drink or round) on the following Sunday. All of the journeymen stocking-knitters would then assemble in the *Herberge* to drink together – at their own cost – with their newly arrived colleague. This was the second portion of the *Geschenk*. The third portion then became due and payable once a journeyman had worked longer than a quarter of a year in Vienna and then either had to or wished to move on. The *Abschenk* or *Ausschenk* (parting gift) was now due – and indeed not in the form of wine but rather in cash: four *Groschen* 'in the bag' as a tramping donation.

All these expenses were financed out of the journeymen's chest or directly by the journeymen themselves. Of course, each journeyman working in Vienna contributed to this fund – one *Kreuzer* per week to be exact. Newly employed journeymen even had to produce an additional initiation fee of six *Kreuzer*, though this was hardly a burdensome amount in relation to a weekly salary of 16 *Kreuzer* plus 7 *Kreuzer* 'beer money' along with room and board in the master's household.

Of course, these provisions cited from the ordinance for journeymen stocking-knitters do not necessarily describe the actual state of affairs. They can rather be regarded as a catalogue of normative demands, the outcome of a bargaining process involving masters, journeymen and governmental authorities – immigrant journeymen were to behave and to be received by their local colleagues in the manner thus depicted. The extent to which actual practice followed the rules established in the ordinance is unknown to us. Nevertheless, this example makes us aware of the great importance of *Geschenk* and *Herberge*, which fulfilled at least three social functions.

First, they were components of a social and communications network. In the *Herberge*, tramping journeymen found material support, information on labour market conditions and many other matters, as well as quite frequently comrades whom they could join on their subsequent travels. *Geschenk* and *Herberge* thus enabled migrant journeymen to survive in an 'honourable' way, even when they failed to find work and had absolutely no income. This kept them from having to seek employment outside their calling, which would have been injurious to their identity as craftsmen, or to eke out a meagre existence in another, less honourable way such as beggar or thief, two occupations into which it would have been easy for men in such circumstances to slip. Moreover, *Geschenk* and *Herberge* constituted institutionalized forms of acceptance into a social milieu and these forms operated in pretty much the same way throughout Central Europe. Even when great distances separated the migrant journeyman from his home town, he was nevertheless able to move in familiar and quite similar social structures.

Second, *Geschenk* and *Herberge* constituted an essential mechanism for the mediation of supply and demand within the labour market. Each journeyman who was granted a form of *Geschenk*, either from an individual master or at the *Herberge*, thereby pledged himself to accept work offered to him. We can interpret the *Geschenk* as a material incentive for tramping journeymen to seek employment with particular masters, workshops or guilds in their craft trade. As is so frequently the case with respect to a donation, *Geschenk* thus had a reciprocal character and was a component within a system of exchange. In many ordinances the obligation to begin work upon acceptance of the *Geschenk* is explicitly stipulated. In the 1675 Vienna ordinance for journeymen stocking-knitters cited above, this is illustrated through the special case of two or more journeymen travelling together who find that there is work available in Vienna for only one of them. If they did not wish 'to be separated, one from the other' or 'one to work without the other', the strangers then had to pay their own tab 'for all that was there imbibed' and were entitled to receive neither accommodation for the night nor any other form of *Geschenk*.[57]

A third function of *Geschenk* and *Herberge* can be described as integration and control. In the ritual of 'welcoming', 'offering greetings' and 'drinking toasts', the tramping journeyman had to conduct himself in accordance with the rules of his trade and to exhibit thereby his affiliation to his craft. These rituals constituted a means of identification in a transregional oral culture. The Sunday round of drinks following the arrival of an immigrant journeyman also constituted the ritual of acceptance within a fraternal community. For the journeymen of the early modern period, and especially in the eighteenth century, ritual demonstrations of unity and solidarity were essential components of the process of accumulation and reproduction of their 'symbolic capital of honour.'[58]

Upon entering the *Herberge*, the immigrant journeymen submitted themselves to the authority of the guild or the journeymen's association. Their presence was now known and they were registered. The thirtieth and final paragraph of the 1675 Viennese ordinance for journeyman stocking-knitters specified that 'a journeymen's book is to be maintained, in which the names ... of all arriving journeymen are to be inscribed'.[59] Indeed, in the eighteenth and nineteenth centuries, books of this kind were kept by the guilds or the journeymen's brotherhoods. Many of them have been preserved. In their enormous detail and diversity, they provide an immense store of source material on the subject of migration by journeymen, on the fluctuations and the consistencies in their working conditions, and a wide range of other questions.[60] Once a journeyman was registered by the guild, it was at least a bit more difficult for him to establish himself illegally in the city in order to work outside the guild and its control.

State regulation of the tramping system

Beginning in the 1720s, governmental authorities in the Habsburg monarchy sought systematically to regulate the system of *Wandern*, *Geschenk* and *Herberge*. In the General Guild Articles for the Bohemian Lands of 1729 a maximum limit for the *Geschenk* was established. In the larger cities it was not to exceed 15 to 20 *Kreuzer*; in smaller towns, six or seven *Kreuzer* or 'one meal consisting of plain fare'. And in tiny villages without a guild, where 'coming up with the *Geschenk*' was unduly burdensome for a lone individual master, a migrant journeyman could no longer claim entitlement to any *Geschenk* and was obliged to move on.[61]

The 1731 General Handicrafts Act and Fundamental Patent (*Handwerksgenerale und Fundamentalpatent*) for the Holy Roman Empire decreed 'that no journeyman [was permitted to receive or demand] more than 15 *Kreuzer*, or 20 *Kreuzer* at the very most, in cash or in the form of food and drink'. An allowance of 15 to 20 *Kreuzer* was by no means an insignificant sum: this approximately corresponded to the weekly pay of a journeyman, who additionally received room and board from his master. The function of the *Geschenk* as a binding commitment to accept work and the role of the *Herberge* were particularly emphasized:

> But if a journeyman wishes only to run from one place to another without taking up work, then nothing is to be given to him ... And each journeyman shall also seek his accommodation in his proper lodging house and refrain from all forms of begging [and], if he has obtained no employment within three days, continue upon his way.[62]

The law then turned its attention, although in what was still a rather vague formulation, to the traditional customs of the journeymen: 'The excesses which have been perpetrated by journeymen as a result of *Schenkung* are ... terminated', that is to say, forbidden.[63]

An essential component of trades legislation in the early eighteenth century consisted of transforming the oral identification of tramping journeymen into a formal procedure. As a means to achieve this end, the General Guild Articles of 1729 and the General Handicrafts Act of 1731 made a *Kundschaft* (letter of reference) mandatory. This document would serve as a form of identification for the migrant journeyman, vouching for the fact that he had successfully completed his apprenticeship and had conducted himself properly with his previous employer.[64] Beginning in 1731, laws mandating these certifying documents were quite promptly written into commercial legislation in most territories of the empire, as well as in Switzerland, as effective instruments of discipline and control of the journeymen.⁶⁵

Precisely what information certifying documents were to contain was stipulated in the General Handicrafts Act of 1731: the journeyman's name, age and place of birth; then a brief description of his physical appearance, in particular his build and hair colour; and finally, the particulars concerning the period of time he had spent at work in the respective city and the addendum that he had 'conducted himself in a diligent, quiet, peaceful and honest manner as befitting a journeyman in the crafts and trades' during this time.⁶⁶ This document was to be issued by the guild and to carry the signature of the 'guild master', a chairman of the corporation elected to serve a specified term. Each journeyman was to receive such a document before his departure from a city and to present it in the *Herberge* or to the guild master upon arrival in the next one. If he did not begin work in a particular city, this fact was to be noted on his document. If, however, he found a job and remained in the city for some time, he was to receive a new *Kundschaft* upon departure.

Early in the 1770s, the Austrian government passed a series of journeymen's ordinances which sought with increasing vigour to dislodge the tradition of *Geschenk* and *Herberge* from the autonomous realm of the journeymen and to subordinate it to control by the authorities. In stark contrast to the journeymen's ordinances of the seventeenth century, these were no longer negotiated and agreed to by masters and journeymen within the guilds. Rather, the governmental authority with jurisdiction over guild matters, the *Commercien-Consess* of Lower Austria, issued a decree to the mayor and the city council of Vienna, ordering them to draw up a journeymen's ordinance and, once it had been reviewed by the state, to put it into effect.

The Articles for the Journeymen Silk and Woollen Stocking Weavers, issued in January 1772 by the mayor and the city council of Vienna, remains to some extent well within the tradition of the journeymen's ordinance of 1675 but shows very distinct signs of the struggle on the part of the authorities against the autonomy of the journeymen.⁶⁷ The highly equivocal concept of *Geschenk* was replaced in the language of state commercial legislation by a new term, *Schlafgeld* (sleeping money or housing stipend), which had been officially conceived as a reference only to its aspect of social support. The ambivalence of 'dispensing', drinking and the ritual acceptance into the community of journeymen was meant thereby to be symbolically dissolved. At the same time, the 1772 ordinance strengthened the connection between *Geschenk* and the new documentary instrument of control, the *Kundschaft*: '... whoever fails to display a *Kundschaft* can receive neither housing stipend nor work ...'. The *Herberge* as well was made mandatory as a point of reception for migrant journeymen but was meant to lose its character as a social centre for all journeymen. Now it should only serve as a lodging house and a house of call:

> When a journeyman silk and woollen stocking weaver migrates here, he shall make his way to the lodging house to put up there and at no other place, immediately displaying his *Kundschaft* to the guild master. With the token he has then received, he is to return to the lodging house thereafter: to wit, he is to receive three *Kreuzer* as housing stipend instead of the journeymen's allowance from the journeymen's chest, as was formerly customary but has been abolished by high and duly-issued ordinance and under the most extreme penalties. The journeyman migrating in the countryside shall either be accommodated free-of-charge by the country master or be given one *Kreuzer* housing stipend by him. When, after three days, no work is to be found for the immigrant journeyman, he is obliged to immediately depart from this place.

As for the journeymen already established locally, it was strictly forbidden 'to assemble [in the lodging house] without giving prior notification to or in the absence of' the municipal craft trades commissioner, to accompany a newly arrived journeyman to the workshop of the master employing him, to provide 'escort' to a departing journeyman or, in general, 'to appropriate funds from the journeymen's chest for the purpose of eating or drinking'.[68]

Autobiographies dating from the early nineteenth century, however, indicate that the practices of 'toasting' and accompanying departing journeymen continued to be quite widespread.[69] The lodging house had retained its place as a centre of communications for the journeymen. In the small crafts, above all, old traditions coexisted with new state regulations. The meaning of these old traditions, however, changed. In the large trades with hundreds or thousands of journeymen, in particular, it would have been impossible strictly to follow the entire body of their ancient customs. Practices and rituals which had once united all journeymen were now practised only in part and only by individual groups. Not all the journeymen of a certain craft would have their welcome drink with the new arrival, but probably only those from the nearby workshops. Not all of them accompanied the departing colleague, but only those who had become personal friends. Instead of a single *Herberge*, three or four inns in different parts of the town fulfilled that function, whether they were officially acknowledged by guilds and authorities or not. Not all the masters would support a tramping journeyman with a proper meal or a donation in cash, but if the first would not then the second probably would. Customs and institutions which were the backbone of the old tramping system did survive well into the nineteenth century, though they did not survive as a proper whole but rather as a fragmentary patchwork.

Notes

1. I would like to thank Melvin Greenwald for his generous help with the translation of this essay into English.
2. G. Jaritz and A. Müller (eds), *Migration in der Feudalgesellschaft*, Frankfurt am Main: Campus Verlag (1988); L.P. Moch, *Moving Europeans. Migration in Western Europe since 1650*, Bloomington: Indiana University Press (1992); C. Lis, J. Lucassen and H. Soly (eds), *Before the Unions. Wage earners and collective action in Europe, 1300-1850, (International Review of Social History*, supplement 2), Cambridge: Cambridge University Press (1993).
3. S. Hochstadt, 'Migration in Preindustrial Germany', *Central European History*, 16, 1983, pp. 195-224, here p. 213.
4. See for instance, among many others, H. Bräuer, *Gesellen im sächsischen Zunfthandwerk des 15. und 16. Jahrhunderts*, Weimar: Böhlau (1989); R.S. Elkar (ed.), *Deutsches Handwerk in Spätmittelalter und Früher Neuzeit*, Göttingen: Schwartz (1983); W. Reininghaus, *Die Entstehung der Gesellengilden im Spätmittelalter*, Wiesbaden: Steiner (1981); R. Reith, *Arbeits- und Lebensweise im städtischen Handwerk. Zur Sozialgeschichte Augsburger Handwerksgesellen im 18. Jahrhundert (1700-1806)*, Göttingen: Schwartz (1988); K. Schulz, *Handwerksgesellen und Lohnarbeiter. Untersuchungen zur oberrheinischen und oberdeutschen Stadtgeschichte des 14.-17. Jahrhunderts*, Sigmaringen: J. Thorbecke (1985).
5. On the economic logic of artisan migration, see K.J. Bade, 'Altes Handwerk, Wanderzwang und Gute Polizey: Gesellenwandern zwischen Zunftökonomie und Gewerbereform', *Vierteljahrschrift für Sozial- und Wirtschaftsgeschichte*, 69, 1982, pp. 1-37; U.-C. Pallach, 'Fonctions de la mobilité artisanale et ouvrière – compagnons, ouvriers et manufacturiers en France at aux Allemagnes (17e-19e siècles)', *Francia*, 11, 1983, pp. 365-406.
6. Commission for Inquiring into the Conditions of the unemployed Hand-Loom-Weavers in the United Kingdom, *Parliamentary Papers*, 1839, 42, p. 116.
7. A typology of artisanal migratory systems is elaborated by R. Reith, 'Arbeitsmigration und Gruppenkultur deutscher Handwerksgesellen im 18. und frühen 19. Jahrhundert', *Scripta Mercaturae*, 23, 1989 (1/2), pp. 1-35.
8. J. Ehmer, 'Räumliche Mobilität im mitteleuropäischen Handwerk', in J. Ehmer, *Soziale Traditionen in Zeiten des Wandels. Arbeiter und Handwerker im 19. Jahrhundert*, Frankfurt am Main: Campus Verlag (1994), pp. 101-29.
9. See R.S. Elkar, 'Schola migrationis. Überlegungen und Thesen zur neuzeitlichen Geschichte der Gesellenwanderung aus der Perspektive quantitativer Untersuchungen', in K. Roth (ed.), *Handwerk in Mittel- und Südosteuropa*, Südosteuropa-Studien 38, Munich: Selbstverlag der Südosteuropa-Gesellschaft (1987), pp. 87-109; Reith, 'Arbeitsmigration'.
10. Studies in the Viennese guild archives are supported by the research project 'Stability and mobility in the Viennese craft guilds, 1740-1860', funded by the Austrian Research Council (P10807 – Soz). For their help and comments, I would like to thank the members of the project team (Heinrich

Berger, Annemarie Steidl, Sigrid Wadauer), Reinhold Reith who supported the project during his stay in Vienna as a fellow of the Lise-Meitner-Program of the Austrian Research Council and Helmut Bräuer as a Visiting Professor in Vienna and Salzburg.
11. R. Sandgruber, *Ökonomie und Politik. Österreichische Wirtschaftsgeschichte vom Mittelalter bis zur Gegenwart*, Vienna: Ueberreuter (1995), p. 107.
12. V. Thiel, 'Gewerbe und Industrie', in Alterthumsverein zu Wien (ed.), *Geschichte der Stadt Wien*, vol. IV, Vienna: Adolf Holzhausen (1911), pp. 411–523, here p. 430.
13. In 1762 in the small Upper Austrian town of Gmunden (1762 population: 1828), there were 66 master artisans per 1000 inhabitants; in the middle-sized Lower Austrian town of Wiener Neustadt (population during the period 1650-1800: ca. 3000–5000), the ratios during the seventeenth and eighteenth centuries oscillated between 61 and 64; in Vienna from 1828 to 1848, between 65 and 68. See J. Ehmer, 'Ökonomischer und sozialer Strukturwandel im Wiener Handwerk – von der industriellen Revolution zur Hochindustrialisierung', in U. Engelhardt (ed.), *Handwerker in der Industrialisierung*, Stuttgart: Klett-Cotta (1984), pp. 78–104; J. Ehmer, 'Vom "alten Handwerk" zum Kleingewerbe. Sozialer und ökonomischer Strukturwandel der kleinen Warenproduktion in Wiener Neustadt', in S. Hahn and K. Flanner (eds), *Die Wienerische Neustadt*, Vienna: Böhlau (1994), pp. 339–68.
14. Thiel, 'Gewerbe und Industrie', p. 431.
15. Cabinetmakers: *Wiener Stadt- und Landesarchiv* (hereafter, WStLA), Alte Registratur 58/1756 (A 1/97); stocking-knitters: WStLA, Innungen/ Urkunden 76/1.
16. This calculation is supported by the analysis of early nineteenth-century census listings: see Ehmer, 'Ökonomischer und sozialer Strukturwandel'.
17. M. Seliger and K. Ucakar, *Wien. Politische Geschichte 1740-1934*, vol. I, 1740–1895 (*Geschichte der Stadt Wien*, vol. I), Vienna: Jugend und Volk (1985), p. 133.
18. See WStLA, Innungen/Urkunden.
19. The large number of guilds reflects the division of labour and the special labour market of an urban metropolis. The small Upper Austrian town of Gmunden in 1762 counted 47 different artisanal occupations, the large German town of Nuremberg about 100; cf. R.S. Elkar, 'Wandernde Gesellen in Oberdeutschland. Quantitative Studien zur Sozialgeschichte des Handwerks vom 17. bis zum 19. Jahrhundert', in Engelhardt, *Handwerker*, pp. 262–93.
20. The largest guilds were tailors (698 master artisans), shoemakers (608), shopkeepers (218), innkeepers (202), beersellers (179), gold- and silversmiths (142), cabinetmakers (142) and cord- and stringmakers (132). Among the smallest guilds were whitesawyers (6), file-cutlers (7), cloth-cutters (7) and bagmakers (8 master artisans). See M. Wagner, 'Kleingewerbe und Handwerk im 18. Jahrhundert', in G. Chaloupek, P. Eigner and M. Wagner, *Wien. Wirtschaftsgeschichte 1740-1938*, vol. I. (*Geschichte der Stadt Wien*, vol. IV), Vienna: Jugend and Volk (1992), p. 155; A. Klose, 'Die wirtschaftliche Lage der bürgerlichen Gewerbe in Wien von 1749-1775', doctoral dissertation, University of Vienna, 1957.
21. K. Pribram, *Geschichte der österreichischen Gewerbepolitik von 1740-*

1860. *Auf Grund der Akten*, vol. 1, 1740-98, Leipzig: Dunker und Humblot (1907), p. 19.
22. Thiel, 'Gewerbe und Industrie', p. 430.
23. M. Fuhrmann, *Historische Beschreibung und kurzgefaßte Nachricht von der Röm. K.u.K. Residenz-Stadt Wien und ihrer Vorstädte*, vol. 1, Vienna: Krausische Buchhandlung (1766), pp. 531-58.
24. M. Bucek, *Geschichte der Seidenfabrikanten Wiens im 18. Jahrhundert (1710-1792)*, Vienna: Verband der Wissenschaftlichen Gesellschaften Oesterreichs (1974); C. Steiner, *Die Bader und Barbiere (Wundärzte) in Wien zur Zeit Maria Theresias (1740-1780)*, Vienna: Verband der Wissenschaftlichen Gesellschaften Oesterreichs (1975).
25. *Österreichisches Staatsarchiv/Hofkammerarchiv* (hereafter, HKA), Commerce/Nr 133 NÖ, Fasz. 63/1.
26. Ibid., 5.3.1770: The Austrian Commercial Authorities (*Commerzien Consessus*) argued 'that, as to the aforementioned guild professions, married journeymen would have to work alongside single journeymen in the same workshop, and it is thus not permitted because, otherwise, no journeyman from the Holy Roman Empire would emigrate into these places, and none who had received his training here would be accepted within the Empire'.
27. For additional information on Central European artisanal marriage patterns, see J. Ehmer, *Heiratsverhalten, Sozialstruktur, ökonomischer Wandel. England und Mitteleuropa in der Formationsperiode des Kapitalismus*, Göttingen: Vandenhoeck und Ruprecht (1991), pp. 185-202.
28. These statistics have been published by Thiel, 'Gewerbe und Industrie', pp. 430ff., and are repeatedly quoted in the subsequent literature. Unfortunately, the original source material was so severely damaged by a fire in the Palace of Justice in 1927 that it can no longer be used.
29. On nineteenth-century immigration to Vienna, especially on the dominance of Bohemian immigrants, see J. Ehmer and H. Fassmann, 'Zur Sozialstruktur von Zuwanderern nach Wien im 19. Jahrhundert', in E. François (ed.), *Immigration et société urbaine en Europe occidentale, XVIe-XXe siècle*, Paris: Recherche sur les civilisations (1985), pp. 31-45.
30. K. Stopp, *Die Handwerkskundschaften mit Ortsansichten. Beschreibender Katalog der Arbeitsattestate wandernder Handwerksgesellen 1731-1830*, vol. 1, Stuttgart: Hiersemann (1982).
31. Ibid., p. 296f. See also U. Puschner, 'Gesellenwandern um 1800. Eine bayerische Fallstudie', in Schriftenreihe der Arbeitsgemeinschaft Alpenländer, *Gewerbliche Migration im Alpenraum. Schriftenreihe der Arbeitsgemeinschaft Alpenländer*, Bozen: Verlagsanstalt Athesia (1991), pp. 83-168.
32. H. Schultz, *Landhandwerk im Übergang vom Feudalismus zum Kapitalismus*, Berlin: Akademie-Verlag (1987), p. 33f., p. 170. On the differences between urban and rural artisans, see F. Lenger, *Sozialgeschichte des deutschen Handwerks seit 1800*, Frankfurt am Main: Suhrkamp (1988), pp. 18-21.
33. E. Schremmer, 'Standortausweitung der Warenproduktion im langfristigen Wirtschaftswachstum. Zur Stadt-Land-Arbeitsteilung im Gewerbe des 18. Jahrhunderts', *Vierteljahrschrift für Sozial- und Wirtschaftsgeschichte*, 59, 1972, pp. 1-40; with respect to rural-urban interrelationships in

Central European craft trades in the early nineteenth century – which show more or less the same mechanism in a changed geographical framework – see Ehmer, 'Räumliche Mobilität', pp. 118ff.
34. M. Mitterauer, 'Zur familienbetrieblichen Struktur im zünftischen Handwerk', in M. Mitterauer, *Grundtypen alteuropäischer Sozialformen*, Stuttgart: Frommann-Holzboog (1979), pp. 98–132, here p. 105.
35. Thiel, 'Gewerbe und Industrie', pp. 432–7.
36. In Vienna in the eighteenth century, for example, there is some evidence of intensive migrational relationships with cities in Saxony in the case of the bagmakers, and with Bohemian cities among stocking-knitters. On inter-urban artisanal migration generally, see Elkar, 'Wandernde Gesellen'; Reith, 'Arbeitsmigration'.
37. H. Zatschek, *550 Jahre jung sein. Die Geschichte eines Handwerks*, Vienna: Verlag für Geschichte und Politik (1958).
38. The statute is reprinted in Bucek, *Geschichte*, XLI–LVII.
39. WStLA, Innungen/Bücher 50/1 (Meisterbuch der Seidenzeugmacher 1711–1779).
40. Bucek, *Geschichte*, 19.
41. Calculation based on the register of the *'bürgerliche* Viennese silk manufacturers', in Bucek, *Geschichte*, III–XLII.
42. WStLA, Innungen/Urkunden 20/1.
43. Calculated on the basis of the listings of owners of all 18 businesses, published by E. Reketzki, 'Das Rauchfangkehrergewerbe in Wien. Seine Entwicklung vom Ende des 16. Jahrhunderts bis ins 19. Jahrhundert, unter Berücksichtigung der übrigen österreichischen Länder', doctoral dissertation, University of Vienna, 1952, pp. 173–97.
44. Ibid., pp. 151ff., 221ff. See also L. Bühler, *Von Schustern, Kaminfegern und Bauleuten. Zur gewerblichen Emigration aus Graubünden bis zum Ersten Weltkrieg*, in *Gewerbliche Migration im Alpenraum*, Bolzano-Bozen: Verlagsanstalt Athesia (1994), pp. 483–509.
45. See A. Grießinger and R. Reith, 'Lehrlinge im deutschen Handwerk des ausgehenden 18. Jahrhunderts. Arbeitsorganisation, Sozialbeziehungen und alltägliche Konflikte', *Zeitschrift für Historische Forschung*, 13 (2), 1986, pp. 149–99.
46. Single travelling apprentices appearing in German lodging house registration books are analysed by Elkar, 'Wandernde Gesellen'. In the *Herberge* for journeymen, there was also frequently a substantial number of apprentices. Indeed, they left behind fewer traces in source materials than did the journeymen. The migration of apprentices in Central Europe still constitutes a gap in research.
47. WStLA, Innungen/Bücher 3/1 (Meisterbuch 1689–1773); Innungen/Bücher 4/1 (Meisterbuch 1776–1822).
48. G. Otruba, 'Wanderpflicht und Handwerksburschenwege im Spiegel Wiener und niederösterreichischer Zunftarchivalien', in *1. Internationales Handwerksgeschichtliches Symposium Veszprem, I, 1978*, Veszprem: Ungarische Akademie der Wissenschaften (1979), pp. 43–9.
49. Stopp, *Die Handwerkskundschaften*, 304ff.
50. WStLA, Alte Registratur 58/1756 (A 1/79: 16.3.1756).
51. See B.-J. Ahn, 'Handwerkstradition und Klassenbildung. Eine sozialgeschichtliche Studie zum Verhältnis von Handwerksmeistern und -gesellen in Frankfurt am Main 1815–1866', doctoral dissertation,

University of Bielefeld, 1991, p. 81; Pallach, 'Fonctions de la mobilité artisanale et ouvrière', p. 382; and Viennese data for the early nineteenth century.
52. See Hochstadt, 'Migration', pp. 217ff.; Reith, *Arbeits- und Lebensweise*; Elkar, 'Wandernde Gesellen'.
53. For further information on Viennese family structures, see J. Ehmer, *Familienstruktur und Arbeitsorganisation im frühindustriellen Wien*, Vienna: Verlag für Geschichte und Politik (1980).
54. K. Schulz, 'Gesellentrinkstuben und Gesellenherbergen im 14./15 und 16. Jahrhundert', in H.-C. Peyer (ed.), *Gastfreundschaft, Taverne und Gasthaus im Mittelalter*, Munich: Oldenbourg (1983), pp. 221–42.
55. *WStLA*, Innungen/Urkunden Articuln, § 76/1.
56. A measure of wine (*Wiener Maß*) was approximately 1.4 litres.
57. Articuln, § 19.
58. A. Grießinger, *Das symbolische Kapital der Ehre. Streikbewegungen und kollektives Bewußtsein deutscher Handwerksgesellen im 18. Jahrhundert*, Frankfurt am Main: Ullstein (1981).
59. Articuln, § 30.
60. *WStLA*, Innungen/Bücher. Analysis of some of the journeymen registers is part of the research project mentioned in note 10 above.
61. *Kaiserl. Koenigl. Theresanisches Gesetzbuch*, vol. I, Vienna: Johann Georg Moessle, k.k. priv. Buchhaendler (1789), p. 223.
62. Ibid., p. 203.
63. Ibid.
64. Ibid., pp. 197, 216.
65. A huge collection of Kundschaften is reprinted in Stopp, *Handwerkskundschaften*.
66. K.K. *Theresanisches Gesetzbuch*, p. 198.
67. *WStLA*, Innungen/Urkunden 76/6.
68. Ibid., §§ 5, 11, 13.
69. See, for instance, G.M. Hofmann (ed.), *Biedermeier auf Walze. Aufzeichnungen und Briefe des Handwerksburschen Johann Eberhard Dewald 1836-1838*, Berlin: Edition Temmen (1936); S. Scholtz Novak (ed.), *Meine Reise 1805-1812. Die Aufzeichnungen des Tuchscherermeisters Johann David Scholtz aus seinen Wanderjahren*, Bremen (1993).

CHAPTER TEN

Artisans in Hungarian towns on the eve of industrialization

Vera Bácskai

For a long time industry in Hungary meant craft industry. At the end of the nineteenth century 44 per cent of the industrial population were still independent masters and another 44 per cent were employed in small workshops, while factory workers made up only 12 per cent of those employed in industry. Moreover, guilds continued to exist until 1872 although their authority to control trade was abolished after 1860. Nevertheless, Hungarian city archives are full of complaints from the end of the eighteenth century about the decline of the crafts and of grievances by guild masters.[1] Indeed, the position of urban artisans was shaken long before the expansion of home industrialization by both the influx of manufactured goods from the western part of the Habsburg Empire and by the development of rural crafts. These were the two factors which undermined the guild artisans' former monopoly on producing industrial goods, a monopoly which previously rested far more on the country's underdeveloped economy than on guild restrictions.

After the expulsion of the Ottomans at the end of the seventeenth century Hungary was a depopulated and devastated country as a consequence of the Turkish wars, the uprisings against Habsburg rule and recurrent epidemics. Organized settlement and mass immigration resulted in the revival of the economy, particularly of agriculture, but the great majority of peasants remained for a long time self-sufficient. However limited the demand for consumer goods had been, it could not be met by artisans, who might have numbered around five or six thousand in the 1720s. (The total population numbered about four million at that time.[2]) Therefore, several benefits were granted to immigrant masters, among others fiscal immunity for 15 years. Once settled, these craftsmen strove to secure their monopoly by organizing or reorganizing the ancient guild system and applied for the renewal of former statutes or for obtaining new charters. About two thousand five hundred guilds were chartered in the eighteenth century, mostly corporations of masters living in privileged towns, but at the end of the century guilds appeared also in market places under seigneurial jurisdiction.[3]

In the countryside only a few essential crafts such as millers, smiths, bootmakers, tailors and wheelwrights were represented. Rural masters were usually less skilled than their urban counterparts and thus did not constitute competition for them. Artisans of the free royal cities met not only the higher demands of the urban population but a great part of the necessities of the countryside as well and thus played a leading role in the economic relationship between towns and their hinterlands.[4] Since civic rights were the precondition for the practice of trade, all masters were burghers (*cives*) with full rights, including participation in the town council and the broader assembly of all burgesses called *electa communitas*. Despite the differences in income and wealth, their socioeconomic standing was more or less stable: the great majority of masters - being burghers, property owners and independent self-employed men - belonged to the upper and middle strata of the urban population.

The growing state control over urban administration impeded the defence of guild monopolies against regulations initiated by the government from the second half of the eighteenth century. Guilds were obliged to present their statutes for verification by the government, in an attempt to abolish such so-called 'abuses' as expensive feasts, high entrance fees, and restrictions on the number of masters. If the charters proved to be legal, guilds obtained new uniform statutes established by the government. This mandatory exchange of statutes provoked the resistance of masters not only because they considered it an infringement of their ancient rights but also because of the high fees they had to pay for the new charters. The change of statutes lasted many years, but finally the guilds obeyed the decree. They also accepted the decisions of higher authorities on admission of applicants formerly refused by both the guild and magistracy. Such interventions occurred often; for example, in Pest about one hundred such cases are known from the period between 1756 and 1780. Overall, however, crafts in the eighteenth century were dominated by guilds.

The survival - or more precisely the late revival - of guild privileges, however fruitful it may have been for the reconstruction of the economy, became by the end of the eighteenth century a serious obstacle to further development. This was not because the restrictive regulations of the guilds hindered large-scale production (as assumed in earlier studies) but rather because of lack of capital and competition from more highly developed western industry. Added to this was the unfavourable economic and customs policy of Vienna, giving preference to the import of industrial goods from the Austrian provinces and agricultural export from Hungary, which blocked industrial expansion and the growth of entrepreneurial spirit in Hungary. Guilds were powerless to prevent the establishment of industrial or commercial enterprises because most of them were founded in the countryside by aristocrats. Those established in towns obtained factory privileges from the government and thus fell outside the sphere of guilds, even though they were hardly more than large workshops. The guilds could do no more than delay their development by lodging protests and complaints. They did not have enough power even to block or control the spread of crafts and guilds to the countryside.

However paradoxical it may seem, the guild system was most detrimental to the masters themselves, for it lent them the illusion of having an omnipotent protector. They did not realize that their welfare and position were due less to their privileges than to the limited demand and restricted market. When at the end of the eighteenth century real market conditions developed, the guild masters were largely unprepared for adapting to the new conditions. Instead they insisted on strict adherence to guild statutes, which became less and less suitable for the elimination of competition, but conserved traditional technologies and restricted innovation. In addition, in the first half of the nineteenth century many guilds - especially the smaller ones - lost their legal basis as they were unable to pay the fees for new charters demanded after 1813. Nevertheless, these 'illegal' guilds continued their activity and neither the local magistracy nor the central government questioned their authority.

Changes in consumer habits began at the end of the eighteenth century and particularly after the boom of the Napoleonic wars that created an enormous demand for agricultural products, stimulating peasant households to join production for the market. The purchasing power of both burghers and peasants increased, particularly in areas favourable to agricultural production and situated along primary roads or waterways.[5] The expanding demand for cheap commodities of good quality could be met neither by the restricted productive capacity of crafts nor by cautious retailers purchasing goods from Viennese merchants at relatively high prices. A great part of the demand was met by merchants procuring goods straight from factories and manufacturers or by itinerant Jewish (and other) pedlars. Some of these were agents of the wholesalers; others independent traders, among them many smugglers. Hence merchants and traders became the most dangerous enemies of the artisans. As many of them were Jews, rivalry often appeared in religious garb. Masters demanded the prohibition of merchants selling goods produced by local artisans - of course in vain.

Obviously that did not mean the slackening of activity against artisan rivals both within the cities and those living in the neighbouring market towns and in the countryside. The competition of the latter became more and more dangerous, as the difference between the structure of urban and rural crafts and that of the countryside diminished as the range of goods supplied by rural crafts increased. Certain crafts existing formerly only in free royal cities, such as carpenters, bricklayers, locksmiths and so on, appeared also in the countryside, siphoning off a part of rural consumers from urban artisans. Thus, contrary to western patterns of proto-industrialization, the spread of industry to the countryside did not contribute to the development of mass production. In fact, it was simply an extension of crafts that led to a more even distribution of artisans and more readily available supplies for the population. That is demonstrated not only by the identity of techniques of production but also by the spread of guild organization in the countryside. From among the thousand new or renewed guild charters issued between 1790 and 1848[6] the majority were granted to masters in market towns and more and more frequently to rural artisans as well. In addition, about a quarter of newly founded guilds were so-called mixed ones, uniting masters of different trades. These were not real professional organizations contributing to the self-identity of a craft but aimed only at defending their rights and local monopoly against urban rivals. Moreover, masters were frequently members of guilds of far-away towns where they used to live. Those guilds were, of course, unable to control the production or defend the interests of their distant members. On the other hand such a practice undermined the solidarity and concord of masters working in the same trade but belonging to different guilds.

Despite the increasing number of guilds in the countryside the proportion of non-member artisans was much higher there than in royal cities. That gave many journeymen an opportunity to become independent in market towns or villages if they were refused admission to a guild in a royal free city. This might have been one of the reasons for the decrease in the average number of journeymen per master from 0.9 in 1782 to 0.7 in 1828,[7] and the cause of labour shortages in periods of prosperity.

Thus both urban and rural craftsmen reacted to the changed situation, and to the ever more acute market competition, by returning to the former mechanism of privileges and monopoly that proved to be adequate for undeveloped market conditions but was ineffective under the new conditions. The conservation of guild monopolies proved successful in the underdeveloped, poor regions but there the masters' prosperity and chances for development were undermined by the limited circle of consumers. In contrast, in dynamically growing cities the artisans' position was shaken by the growth of local consumers whose demands could not be met by the limited output of the small workshops. In 1828 only about one-third of masters employed journeymen continuously or temporarily. Most of these - about 75 per cent -

employed only one, another 20 per cent three to five journeymen and there were only 173 masters (1.7 per cent of the total) who employed more than five. The great majority of the latter belonged to the construction industry. The proportion of masters employing journeymen varied according to the type of town. In the capital more than half of the masters employed journeymen, while in towns with accentuated agricultural character less then one-third of them worked with skilled help, and only a quarter of them did so in the small towns.

As already mentioned, the highest number of journeymen was recorded in the construction trades. Most of the labour force was employed by the carpenters and bricklayers; in these two crafts 71 masters each employed more then ten journeymen in 1828. Carpenters and bricklayers with 20 to 30 journeymen were not an exception, and in the largest enterprises the number of skilled employees exceeded 50. Besides journeymen some unskilled labourers were also employed in the building trades; thus the masters of the building trades were the ones who succeeded in establishing large enterprises within the guild organization. They belonged to the small group of wealthy craftsmen, and many of them or their descendants became members of the bourgeois elite in the second half of the nineteenth century.

In other industrial branches larger workshops were to be found only in a few cities: more than two-thirds of masters employing more than two journeymen were concentrated in seven towns, and 40 per cent of them were employed in the capital. Workshops with more then five journeymen existed only in 15 towns. There were altogether nine masters employing more than ten (five tailors, three cabinetmakers, one glovemaker), and five of them lived in the capital. Besides the building industry, journeymen in significant numbers were employed by bakers, butchers, cabinetmakers, locksmiths and smiths. In these crafts the number of journeymen per master ranged from 0.9 to 1.1, double the average of 0.5 in other crafts.[8] Even if they belonged to the most successful and prosperous masters, the production of their workshops did not surpass the limits of craft industry and therefore did not require separation from the guilds. There were, of course, complaints against them by fellow masters but they were easily counterbalanced by their influence on city councils, not to mention their leading position in the corporations themselves where the better-off masters usually held the post of warden.

The limited number of large-scale urban industrial enterprises were founded mainly by merchants connecting industrial activity with wholesale trade. First of all they were those who had enough capital to establish manufactures. They had nationwide – and often even foreign – business connections and in consequence influential protectors in the government, who paved their way to obtain the factory privilege which exempted them from the control of guilds, if not from their interference.

Among industrial entrepreneurs of the early nineteenth century there were but few craftsmen. Many of them succeeded in expanding their workshop by combining industry with trade. As a first step they usually gave up craft for commercial activity and after accumulating some capital returned to industry as manufacturers. This type of career characterized the most prominent textile-mill owner families of Pest - the Goldbergers and the Valeros - the owner of a hat-manufacturers, Karczag, and some tobacco-manufacturers.[9] It is symptomatic that industrial investments were rather ephemeral 'adventures' and former craft masters preferred commercial or financial activity to industry, just as their merchant counterparts. A typical example of this was the career of a tailor in Pest called Libasinszky. His trade was plied by numerous masters who usually worked alone without journeymen and belonged generally to the group of low-income taxpayers. Libasinszky was a newcomer, an immigrant from Silesia in the first decade of the nineteenth century. His workshop, compared to those of his fellows, might be considered under contemporary conditions as a garment factory: in 1818 he employed ten journeymen and seems to have also engaged several masters as outworkers. Moreover, he also employed a book-keeper, a rather exceptional feature among artisans. Libasinszky belonged to the wealthy upper strata of masters: he owned two houses: one in the city, another in a newly established quarter of the town, preferred by wholesalers and financiers. This latter, a three-storeyed house, was built by the famous architect Joseph Hild and brought an annual rental income of 2000 florins, an enormous sum compared to the 244 florins average rental income of artisans.

In the 1830s he apparently gave up his workshop, as in his probate inventory from 1837 there is no mention of cloth or fabric. The items listed in the inventory prove that he changed craft to trade in agricultural produce, and owned a coffee-house that he let by lease. His fortune amounted to a total of 180 000 florins, encumbered with debts of 72 000 florins.[10] The size of this fortune would have allowed industrial investment but, like many other wealthy merchants, he used his success and wealth to raise his social standing and the promotion of his heirs in the social hierarchy. None of his sons followed the trade of his father: one became an officer, one a civil servant, the third a wholesaler and banker. His daughter married a lawyer originating from an artisan family, who soon became a member of the city council. Thus one can conclude that the few successful and well-to-do craftsmen preferred increased social standing to the augmentation of profit and wealth.

From among the few lesser factories founded by former craft masters there were several leather-factories established by tanners, a furniture-factory of a cabinetmaker, factories of coppersmiths and manufactories producing coaches. Most of them were small enterprises and existed for only a short time. These manufactories and factories, together with the limited number of workshops with several journeymen, were the exceptions. Lacking eighteenth-century nationwide comparative data, it is hard to draw general conclusions about changes in the number and proportion of journeymen. At any rate, in the privileged towns both diminished between 1782 and 1828. While the number of masters increased by 24 per cent, that of the journeymen decreased by 5 per cent and the average number of journeymen per workshop sank from 0.9 to 0.7. In consequence the number of artisans per 1000 inhabitants fell from 88 to 66,[11] a ratio far less than in West European towns or in cities in the western part of the Habsburg Empire. Presumably the spread of crafts to the countryside enabled a large number of journeymen to set up shop in market towns and villages.

In any case, small-scale urban craft production was not able to meet the demands of local consumers or those of the countryside. Under constraint by the government, aiming at a certain liberalization of trade, and by the growing demand for cheap commodities, town councils were forced to make more and more concessions to free trade. Ever more merchants and traders were admitted and councils permitted craftsmen to operate outside the guilds, including journeymen. This trend is obvious from the growing number of taxpaying journeymen particularly in larger towns, such as the capital. While in 1790 journeymen in Pest amounted to 12 per cent of all inhabitants paying income tax and 27 per cent of all taxpaying craftsmen, in 1840 a quarter of all taxpayers and half the artisans were not masters. A similar but less accentuated trend can be observed in Buda: while in 1793 the proportion of assessed journeymen made up 3 per cent of all taxpayers and 9 per cent of all artisans, in 1840 the ratios were 11 and 28 per cent respectively.[12] The number of outsiders grew also, evidenced by the appearance of craftsmen working in new products, fashionable goods that had never been produced in the guilds.

From the beginning of the nineteenth century onwards crafts became more and more a 'local' function, an omnipresent branch of the economy. Consequently, the masters played a less important role in basic urban functions and in the economic relations of town and countryside. They lost their former relatively stable position in the higher strata of urban society. All this weakened their self-esteem.

Income tax assessments show a decline of craft profitability which varied according to branch, part of the country and types of town. In the largest city, Pest, the comparison of income tax assessments from the late eighteenth with those of the mid-nineteenth century indicates a trend, though not a linear one, towards the growth of the proportion of masters paying the lowest taxes. In 1784 they accounted for only a quarter of all artisans; by 1840 their proportion had reached 82 per cent while the proportion of masters paying the highest taxes gradually decreased from 16 to 6 per cent. At the end of the eighteenth century the proportion of artisans in the highest tax bracket surpassed that of the total of taxpayers; by 1840 it was much lower. While at the end of the eighteenth century the numbers and proportions of artisans and merchants in the highest tax bracket were equal (both 147 people, 45 per cent), by 1840 merchants together with manufacturers made up 66 per cent of this bracket, while the proportion of highly taxed artisans sank to 16.[13]

A similar but not so marked trend can be observed in the second largest town, Debrecen in Eastern Hungary, where rural consumers had less purchasing power. Here, too, the proportion of artisans paying the highest taxes decreased sharply; however, almost half of the craftsmen preserved their position in the middle ranks of taxpayers[14] (see Table 10.1).

The relatively favourable position of craftsmen in Debrecen was perhaps due to the very small number of artisans in the countryside, allowing crafts to preserve an important role in the economic relations of town and country. That was a common feature of towns situated in the underdeveloped north-eastern regions. A further comparison demonstrates the impact of the development of rural crafts on the position of urban masters. Towns included in Table 10.2 were riverside settlements: Esztergom and Györ in the western part of the country were on the Danube, while Szeged in the south enjoyed the benefits of navigation on the Tisza. In their hinterlands lived a great number of artisans practising various crafts. The development of craft industry surpassed the national average. Table 10.2 shows that the distribution of masters in the tax brackets in these three towns differed from that in Pest.[15]

Table 10.1 Distribution of masters in Pest and Debrecen according to income tax paid (%)

Tax (florins)	Pest			Debrecen		
	1797	1806	1837	1797	1805	1839
Up to 5 fl.	77.9	59.8	74.1	56.9	45.4	44.5
5–10 fl.	10.5	29.9	20.2	34.2	42.4	46.7
Over 10 fl.	11.6	10.3	5.7	8.9	12.1	8.8

Table 10.2 Distribution of masters in Esztergom, Gyôr and Szeged according to income tax paid in the 1840s (%)

Tax (florins)	Esztergom	Gyôr	Szeged
Up to 5 fl.	87	88	98
5–10 fl.	10	7	1
Over 10 fl.	2	5	1

The distinct distribution amongst the different tax brackets of merchants in these towns demonstrates clearly that income from commerce generally surpassed that from crafts (Table 10.3). A change can be observed in the composition of the most profitable trades. The most populous branch, the clothing industry which once ranked high, was ousted with the exception of Debrecen and replaced by artisans of the food industry, building trades and metalworking crafts. The highest taxes were paid by bricklayers, butchers, glaziers, blacksmiths, chimney-sweeps, locksmiths, brewers, coppersmiths and hatters. These were also the crafts where most journeymen were employed and those in which masters had an average rental income exceeding that of other craftsmen: they usually owned more than one house or if only one, it was larger than the average and contained several flats.

The changes in profitability and prosperity of individual crafts are well demonstrated by changes in the tax assessments of Pest. Some 275 masters in 17 trades were assessed higher in comparison to their eighteenth-century assessments, while taxes of 553 masters in 31 trades were lower; in about half the cases the decrease was significant. Thus development and decline of crafts were rather selective, but decline was overall the more characteristic experience.

This fact was most symptomatic in the capital. In Buda one-third of the masters paid as much tax as, or less tax than, the assessed

Table 10.3 Distribution of merchants in Pest, Esztergom, Gyôr and Debrecen according to income tax paid in the 1840s (%)

Tax (florins)	Pest	Esztergom	Gyôr	Debrecen
Up to 5 fl.	56	70	47	13
5–10 fl.	14	16	23	49
Over 10 fl.	30	14	30	38

journeymen, and the tax imposed on 10 per cent of them equalled the minimal tax imposed on day labourers.[16] The number of property-owning masters diminished as well. While formerly almost all of them lived in their own house and owned some land, mostly vineyards, in 1828 only about 60 per cent had a house, and one-third had land. The proportion of property-owners barely exceeded the share of property-owners in the urban population as a whole; a higher proportion of artisans owned property than wage-earners or merchants but a lower percentage than the professionals, shippers or hauliers.[17]

The lower middle-class position of craftsmen was also reflected in their humble housing conditions and simple lifestyle. The overwhelming majority of them owned or rented flats consisting of one or two rooms,[18] furnished with cheap and simple pieces of strict necessity (in many cases the number of beds was substantially smaller than that of family members). Only every fifth household employed maidservants, much less than merchants or professionals, and the number of employed journeymen living with the family decreased.[19]

The impoverishment of a large proportion of masters and the growing gap between the bulk of masters and the few rich is suggested also by data on wealth patterns. A significant decrease of family fortunes can be observed in several towns after the first decades of the nineteenth century. An investigation of probate inventories taken between 1790 and 1848 in two towns, Gyôr and Sopron, proves that the decline began in the 1820s: the average amount of property fell while total debts and their proportion of the whole property both increased.

The average value of property owned by artisans in Gyôr ranked third in the hierarchy of wealth at the turn of century. Later they were preceded not only by merchants and caterers but also by professionals and clerks. By the 1830s and 1840s they sank to fourth or fifth place. In Sopron they were ranked lower, holding fifth or sixth place from the beginning, preceded not only by merchants, professionals, caterers and shippers, but, from the 1820s onward, by vine-dressers as well. The proportion of debts to total property increased from 5 per cent to 46 per cent in Gyôr and from 39 per cent to 56 per cent in Sopron, exceeding the average for the population as a whole. Ever fewer masters appear in the class of major property-holders: whereas between 1790 and 1820 four artisans belonged to the ten richest citizens of Gyôr, in the 1820s only two of them were to be found in this class and they ranked eighth and tenth. In the 1840s no artisan belonged to the class of citizens of major wealth. In Sopron, similarly, craftsmen disappear from the class of wealthiest families, ousted not by merchants but by vine-dressers.

The structure of property shows that although artisans were keen to live in their own homes, they were not particularly concerned with the comfort or elegance of the apartments. That is proved both by the very low value and the simple composition of furnishing, house utensils and clothing. While the proportion of the value of real estate to the total increased in Gyôr between 1790 and 1848 from 31 to 67 per cent (surpassing the urban average of 42 per cent), the ratio of the value of movables sank from 44 to 23 per cent, far less than the urban average of 32 per cent. In Sopron in the same period the proportion of the value of real estate - apart from a temporary increase in the 1810s to 1820s - amounted to 50 per cent of the total, equalling the urban average, while the ratio of movables decreased from 15 to 12 per cent.[20]

Similar conclusions can be drawn from the probate inventories taken in Pest in the years 1840-42. Although the very limited number of them does not allow statistical generalizations, they show a similar trend. In addition, they offer insights into the lifestyle and housing conditions of the few successful masters.[21] In Hungarian towns probate inventories were usually drawn up for families with property, particularly in cases of heirs under age; thus inventories reflect mostly the wealth and housing condition of well-to-do citizens. Altogether 150 inventories are known from the period 1840-42, accounting for 2 per cent of all those deceased in these years. Consequently they cannot be considered even as a representative sample of well-off inhabitants. After 48 deceased of unknown occupation and status, the second largest group consists of 44 craftsmen; the number of merchants, caterers and professionals and officers was each more than ten; other occupations occurred in small numbers.

After deducting the debts, the net sum of legacies ranged from a few florins to 367 000 florins. In 22 cases - including the legacies of seven craftsmen - the amount of debts surpassed the total. Most of the legacies ranged from 1000 to 5000 florins; about two-thirds of the deceased masters belonged in this bracket, while the same ratio of merchants left an inheritance of 5-10 000 florins. The group of the ten wealthiest families (their legacies ranged from 40 000 to 368 000 florins) was composed of the families of three merchants, three craftsmen, two clerks, a coffee-house owner and a gardener. Thus masters were clearly under-represented, as only 7 per cent of them belonged to the group of the richest opposed to a quarter of merchants and about a fifth of professionals and clerks.

Details of business activity, family background, family strategy and composition of property of these three richest artisan families offer an insight into the conditions and factors of economic success and social ascension of masters whose lifestyle, in spite of their membership in the corporation, differed from that of the average guild member. Two of the three masters practised the most profitable crafts. The largest inheritance was left by the wife of a smith called Ferenc Unger; she was the daughter of a wealthy stonemason. Husband and wife were both descendants of wealthy and prestigious artisan families settled in the town since the mid-eighteenth century. The smith employed regularly some five or six journeymen. His nephew and his brother-in-law were ironmongers (the latter was raised to the nobility), and he may also have been involved in that trade. Ninety per cent of the property, amounting to 368 000 florins, consisted of six houses, arable land and a vineyard with a summer residence. The value of jewels and silverware amounted to more than 4000 florins, a sum not matched in any other estate, and the 8000 florins in cash was a rarity too. Their spacious flat was furnished with valuable pieces made of fine hardwood and there were several reception rooms where guests would presumably be received.

The careers of their children suggest that wealth and luxurious housing conditions were instruments in the social ascension of the family. Neither of his sons continued the occupation of the father, nor did they choose to become a merchant, though they would have had valuable support for that in the wider family. The elder son became an officer, the younger a lawyer and the two daughters married the brothers of a famous composer, presumably professionals or officers. Even though the industrial enterprise of Unger did not provoke conflicts with the guild, his wealth, living standards, family strategy and social aspirations were very different from those of typical guild masters, and closer to the behaviour of merchants and industrialists of entrepreneurial spirit.

The third largest inheritance was left by the widow of a tanner. He had come from a wealthy and ancient artisan family and the widow was from a well established family of butchers. The legacy amounted to 80 000 florins and consisted of a house and a tannery which she owned jointly with her brother-in-law. The high value of the workshop suggests large-scale production; the fact that the brother-in-law was called an 'industrialist' (*Fabrikant*) implies that the tannery must have been free from the control of the guild.

While the family background and the practice of profitable crafts played a great role in the rise of the two masters mentioned above, the third craftsman, a tailor who ranked sixth, leaving an estate of 58 000 florins, was an immigrant. He acquired burgess status six years later in 1822 by which time he employed seven journeymen and sold his products in a shop. The inventory listed a number of fine garments to the value of 2300 florins. It was not only his relatively large-scale production but also the combination of craft and trade which set him apart from most other tailors. He owned two houses to the value of 55 000 florins, but his housing conditions were very modest. Perhaps as a new settler he was more concerned with the establishment of his business and the future rental income of his seven children.

In contrast to the few successful careers such as these, however, an increasing proportion of artisans fell into the lower middle class, or even into the lower strata of urban society. In consequence, their social standing declined as well. Masters sensed the deterioration of their situation but did not realize that it was part of an irresistible long-term process. They believed that the precarious financial situation was only transitory, and that equilibrium could be restored by regulating competition. They also hoped that craftsmen's reputations could be retrieved by strengthening the formerly prestigious organizations, such as guilds and burgher rights, and through these regain their influence on urban government. Thus, once again, they cherished illusions despite their everyday experience which proved the inefficiency and powerlessness of these institutions now deprived of their original character.

As mentioned before, by the first half of the nineteenth century guilds had lost their ability to safeguard the economic interests of their members or to preserve the feeling of professional or religious community. Journeymen, who tended not to be lodged at the master's house but, as married men, founded their own households, became wage-earners.[22] The former patriarchal artisan household was about to be dissolved. Increasing numbers of once independent masters were forced to close their shops and seek employment with their more successful fellows, or - particularly in the garment industry - become outworkers for commercial firms.

The guild no longer united all artisans in a trade, as the number of outsiders gradually increased. Moreover, competition caused increasing numbers of disputes and accusations, leading frequently to open scandals both inside the guilds and between different organizations. Statutes were disregarded by guild members themselves: besides their own products (and repairs) they began to sell manufactures, an activity strictly prohibited by their rules.

The increasing tension and animosity were a block to sociability within the corporation. In reality it seems doubtful whether sociability, solidarity, common rituals - all those idyllic pictures one finds in many works on guilds - were a reality or only a myth derived from reading the meticulous regulations in guild statutes. Lacking adequate sources it is hard to answer this question but some indirect data (for example the rare intermarriage between families of the same craft despite the broad endogamy of artisans in general) indicate the dissolution of the former close relations between guild members.[23] Finally, guild autonomy had long before ceased to exist. Each corporation was subject to a designated alderman who had to be present at each assembly and control the guilds.

Still, the masters expected support from the city council and showered complaints and grievances upon it against artisans licensed by the council or engaged in illegal trade.

Protests were mostly fruitless, for either the accused lodged an appeal or, to avoid annoying discussions, the council itself passed such protests on to governmental offices which rarely took the side of the guild members. The guilds' confidence in the town council was based on past experience when most aldermen came from artisan families or were active craftsmen. However, nineteenth-century aldermen, though frequently related to artisan families, became a separate group of trained office-holders, increasingly subject to the government. Their training, position, lifestyle and interests became assimilated to those of the professionals and lower nobility, the more so as a part of them was raised to the nobility and/or connected with it by marriage. Influenced by loyalty to the government and to some extent by liberalism, they could rarely act on behalf of the masters.

As they became aware of this situation, craftsmen cherished the hope of exerting pressure on the council by making use of the privileges that went with their civic rights. That plan again proved to be an illusion because by the nineteenth century the once meaningful concept of burgess rights had become a mere legal category. The citizens who in the middle ages had fulfilled the most important role in the economic life of towns, who participated in the administration of the towns and who alone enjoyed urban liberties had become a small fraction of the urban population without any significant privileges. Their influence on the administration of the town was minimal; the economic role and wealth of craftsmen and shopkeepers lagged far behind those of many a newcomer without civic rights, making a good profit from trade.

The number of burgesses had previously been limited by city councils to protect the monopoly of guild members. By the beginning of the nineteenth century no such regulations were needed because civic status was no longer the precondition of pursuing industry and commerce; hence, its acquisition meant only expenses rather than benefits. Consequently, fewer and fewer people strove to obtain it. The proportion and composition of burgesses varied according to the conditions of different types of towns with different patterns of development but, for example, in 1828 only 60 per cent of masters living in free royal cities were burghers. Usually masters belonging to the most profitable crafts were more eager to obtain civic rights, for the proportion of burghers among them exceeded 70 per cent.

There were, however, exceptions. The example of Pest demonstrates that it was precisely the wealthy inhabitants with an enterprising spirit who were least interested in legal status. By this time the acquisition of civic status was important mainly for those who wished to take root in the town: primarily for immigrants, lesser masters, retailers, for people of modest wealth. Thus in Pest citizenship became an institution serving social integration and upward social mobility for newcomers. In more traditional towns with a less differentiated occupational composition, as for example Buda, citizens became more and more an exclusive group of local-born artisans and vine-dressers. Thus burgess status became an instrument against strangers and newcomers.

Even though citizenship lost much of its former prestige and benefits, it was the prerequisite for becoming a member of the external council, the *electa communitas* (*Wahlbürgerschaft*), a corporation designed to control the council's activity, but with very little real authority. Even though the corporation controlled urban finances and had the right to make proposals on other aspects of city administration, it was frequently ignored. Its remaining function was to elect members to the council. However, since aldermen were elected for life from among burgesses nominated by the town council, elections could only fill places that fell vacant by death. Still, to obtain some influence on town administration through the external council the first step was to acquire civic rights.

In the absence of detailed research, this process can be demonstrated only for the case of Pest, the growth and social composition of which was, no doubt, somewhat special. While at the end of the eighteenth century and in the first two decades of the nineteenth the proportion of artisans among new citizens was about 37–39 per cent, from the 1820s it increased to 50 per cent and this ratio was preserved until 1848. In the same period they secured half the places on the wider council. Even though 46 per cent of the wider council around 1840 consisted of masters belonging to the most profitable trades (in contrast to the late eighteenth century, when they reached only one-third), the majority of this body still came from craftsmen in the clothing industries. As we know, the profitability of that most populous branch of industry was on the decline. Hence the majority, however slim, belonged to the most conservative section of artisans, in both economic and political respects.[24]

The strong presence in the wider council was an illusory success as that body was not able to influence the decisions of the town council or even prevent it from implementing government orders detrimental to the crafts. Nevertheless, they were able to delay important decisions by endless protests against 'infringement of ancient privileges'. That was their tactic in dealing with such questions as the building of a permanent bridge, the expansion of suffrage and the democratization of town administration. The reason for their success was that aldermen and members of the external council were in agreement on resisting liberal reforms.

Until 1830 the nomination of deputies to the diet was monopolized by the town council. Thereafter, in order to demonstrate their readiness for democratic reforms, members of the external council were also involved. They tried to prevent a further extension of suffrage to all citizens, but in vain. In the 1840s they had to accept being supplemented by a small electoral group elected by the whole community of citizens.

In order to ensure support for their own interests, artisans were very active in elections during the 1840s. While about 25 per cent of merchants and 30 per cent of professionals participated in the urban elections, about 50 per cent of artisans went to the polls. A similar activity is apparent in the parliamentary and municipal elections following the revolution of 1848 when they succeeded in defeating several liberal professionals. The revival of political activity by artisans, together with labourers' demonstrations and movements, no doubt contributed to the fact that the government, in order to win their support, changed its mind and did not abolish guilds, but confined itself to essential reforms undermining the monopoly of guild masters. In the 1850s a new corporation, the Chamber of Commerce and Industry, was founded which was somewhat distrusted by artisans because of the determining influence of merchants on the board.

In the second half of the nineteenth century artisans, or at least some of them, succeeded in adapting to the requirements of the market economy and to the developing capitalist conditions. Some of them even became factory-owners by transforming their workshops into industrial enterprises and thus advanced to the upper middle class. Despite bankruptcy and pauperization of many artisans the increasing number of independent masters and the preponderance of small workshops until the end of the nineteenth century prove the success of this learning process. No doubt there were many newcomers among them, as only a third of the new workshop-owners can be identified as former guild members or descendants of masters active in the first half of the century. This means that the opportunities afforded by free trade were grasped partly by artisans formerly excluded from the guilds and partly by newcomers free of traditional corporate spirit. In the beginning the increase of independent artisans exceeded that of employees,[25] but this trend changed at the end of the century with the rapid growth of industry.

The statement that industrialization ruined craft industry has long been discredited. What in fact happened was a restructuring of small-scale industry both in countries of advanced industrialization and elsewhere. In economies hit by the mass production of Western Europe this happened long before the development of their home industry. Growth and decline were, however, selective: while some crafts suffered from the influx of mass-produced goods, others, such as service industries and mechanical and building trades, gained from it. Craft industry was not ruined by industrialization but gradually changed into a branch of production of secondary and local importance. It was the social standing of artisans which suffered most. This was manifested not only in the decline of the majority into the lower middle class and of a considerable part into the class of skilled factory-workers, but also in the change in the social status of successful masters. The small group of rich artisans who invested their capital in industry, or in commercial and financial enterprise, rose to the upper middle class and became part of the bourgeoisie. The split fragmented the occupational group which, in spite of division of wealth, prestige and different levels of authority, had once constituted a more or less uniform estate.

Notes

1. Many of them quoted by K. Dóka, *A pest-budai céhes ipar válsága 1840-1872* (Crisis of craft industry in Pest-Buda 1840-1872), Budapest: Akadémiai Kiadó (1979); G. Eperjessy, *A szabad királyi városok kézmûvesipara a reformkori Magyarországon* (Craft industry in the Hungarian free royal towns in the first half of the nineteenth century), Budapest: Akadémiai Kiadó (1988).
2. G. Ember and G. Heckenast (eds), *Magyarország története 1686-1790* (History of Hungary 1686-1790), Budapest: Akadémiai Kiadó (1989), p. 25.
3. G. Eperjessy, *Mezôvárosi és falusi céhek az Alföldön és a Dunántúlon, 1686-1848* (Guilds in market towns and villages in the Transdanubian region and in the Great Hungarian Plain), Budapest: Akadémiai Kiadó (1967).
4. This is the reason for the great attention paid to artisans and guilds by local historians. A selection of recent studies in foreign languages is published in *Internationales Handwerkgeschichtliches Symposium Veszprém, 1978*, I, Veszprém (1979); II/1-2, Veszprém (1983); III/1-2, Veszprém (1986); IV, Veszprém (1995).
5. V. Bácskai, *Towns and urban society in early nineteenth century Hungary*, Budapest: Akadémiai Kiadó (1989), pp. 32-48.
6. I. Éri, L. Nagy and P. Nagybákay, *A magyarországi céhes kézmûvesipar forrásanyagának katasztere* (Register of sources relating to crafts and guilds in Hungary), Budapest (1975), pp. 335-44.
7. V. Bácskai and L. Nagy, *Piackörzetek, piacközpontok és városok Magyarországon 1828-ban* (Market areas, market-centres and towns in Hungary in 1828), Budapest: Akadémiai Kiadó (1984), p. 169.
8. Bácskai, *Towns and urban society*, pp. 62-3.
9. V. Bácskai, *A vállalkozók elôfutárai* (The forerunners of entrepreneurs), Budapest: Magvetô (1989), pp. 99-101, 104-9.
10. Ibid., pp. 40-42.

11. Bácskai and Nagy, *Piackörzetek*, pp. 176-8.
12. V. Bácskai, 'Társadalmi vâltozások Pesten az 1830-1840-es években' (Social changes in Pest in the 1830s-1840s), in T. Faragó (ed.), *Pest-budai árvíz 1838*, Budapest: Fôvárosi Szabō Ervin Könyvtár (1988), pp. 213, 218.
13. Ibid., pp. 212-19.
14. Bácskai, *Towns and urban society*, pp. 66-8.
15. Esztergom: ibid., p. 86; Gyôr: P. Balázs, *Gyôr a feudalizmus bomlása és a polgári forradalom idején* (Gyôr in the period of decomposition of feudalism and of the bourgeois revolution), Budapest: Akadémiai Kiadó (1980), p. 66; Szeged: J. Farkas (ed.), *Szeged története 2. 1686-1849* (History of Szeged, vol. 2. 1686-1849), Szeged (1985), p. 373.
16. Bácskai, 'Társadalmi változások', p. 219.
17. Bácskai, *Towns and urban society*, p. 104.
18. Ibid., pp. 74-5.
19. Ibid., pp. 62-4, 78.
20. V. Bácskai, 'A polgári vagyon nagysága és szerkezete a XIX. század elsô felében' (Structure and value of middle-class property in the first half of the nineteenth century), in S. Farkas (ed.), *Gyôri Tanulmányok*, vol. 13, Gyôr: Gyôr Megyei Jogú Város Polgármesteri Hivatal (1993), pp. 39-57.
21. Bácskai, 'Társadalmi változások Pesten', pp. 220-27.
22. In Vienna, in contrast, the number of journeymen lodged in the master's house increased: see J. Ehmer, 'Ökonomischer und sozialer Strukturwandel im Wiener Handwerk von der industriellen Revolution zur Hochindustrialisierung', in U. Engelhardt (ed.), *Handwerk in der Industrialisierung*, Stuttgart: Klett-Cotta (1984), p. 94.
23. The diary of a barber's assistant reports mainly on journeymen's drinking-bouts in local taverns, sometimes in the presence of the master. Himself a drunkard and idler, the assistant became acquainted with masters and guilds of several towns in the course of his wanderings. In his diary there is no trace of the idyllic conditions. Maybe this is to be ascribed to his deviant personality, but with respect to the solidarity of masters, it is rather thought-provoking that one of his employers, in order to get rid of him, sent him to a distant fellow master with a very warm recommendation. K. Francsics, *Kis kamorámban gyertyát gyujték* (I lit a candle in my little closet), Budapest: Magvetö Könyvkiadó (1973).
24. Bácskai, 'Társadalmi változások', pp. 228-31. The occupational composition of the *electa communitas* has not yet been the subject of research. The 1828 census indicated the composition of the elected town bodies only in Buda and Szeged. In Buda 13 out of 27 artisan members, i.e. 48 per cent, were representatives of the best-off masters; in Szeged five out of 16 (33 per cent): Bácskai, *Towns and urban society*, p. 70.
25. A fluctuation in the increase of employers and employees and their changing proportion can be observed in Germany and Austria as well, though the Hungarian process was somewhat postponed. See J. Ehmer, 'Ökonomischer und sozialer Strukturwandel'; G. Schmoller, *Zur Geschichte der deutschen Kleingewerbe*, Halle (1870); M. Birnbaum, *Das Münchener Handwerk im 19. Jahrhundert (1799-1868)*, Munich: Handelskammer für Oberbayern (1984).

CHAPTER ELEVEN

Urban renovation and changes in artisans' activities: the Parisian *fabrique* in the Arts et Métiers quarter during the Second Empire*

Florence Bourillon

During the Second Empire, Paris was profoundly remodelled and transformed. The imperial authorities - Napoleon III himself, along with his prefects of the Seine department, Berger and then Haussmann - took the decisions about these changes. Haussmann remained at the post of prefect from 1853 until 1870 and both his detractors and his enthusiastic supporters became used to applying his name to describe the urban renovation of the French capital. Major works were undertaken on three occasions during the Second Empire: in 1852, in 1858 and then towards the middle of the 1860s. Each of these campaigns comprised a series of operations, from the opening up of new thoroughfares to the construction of new hospitals, local town halls for the *arrondissements* of Paris, the creation of parks, market halls and so on. Each campaign constituted a 'network', characterized by a certain autonomy in the way it was financed and by the specific aims underlying the way it evolved over time.[1] The first campaign, launched in 1852, was primarily intended to develop the Île de la Cité, opening it up by means of new boulevards, Sébastopol and Strasbourg to the north, the boulevard Saint-Michel heading south and the east-west extension of the rue de Rivoli, parallel to the Seine, on the right bank. The second major campaign was planned from 1858 in preparation for the annexation of the peripheral communes surrounding the old capital - the communes of the 'immediate suburbs' incorporated into the enlarged city in 1860. From this point of view, the second campaign was concerned to facilitate communications between the different quarters inside the city. The main, city-wide 'cross-roads' having been established, this campaign launched new, diagonally oriented thoroughfares, radiating, in the west, from the place de l'Étoile and, in the east, from today's place de la Nation. Cutting across existing

---

*Translated from the French by Paul Smith.

streets, these new diagonals created a series of secondary *étoiles* or circuses. The third campaign had similar aims to the second, finishing off a certain number of the operations already begun, particularly in the 'annexed' suburban communes, but also in the centre of the city, notably with the creation of the boulevard Saint-Germain on the left bank.[2]

Two apparently contradictory logics were at work in these programmes (Figure 11.1). On the one hand there was a centralizing tendency, developing the administrative infrastructures in the Cité, whilst on the other hand a decentralizing movement sought to develop the outer *arrondissements*, giving new administrative functions to their local town halls. But in fact these apparently contradictory tendencies served complementary aims. Haussmannian town planning envisaged a distribution of functions on the level of the whole agglomeration. Seen

11.1 Major works in Paris during the Second Empire

in this way, the administrative centre was balanced by corresponding zones of housing and by neighbourhoods of commercial exchange around the main railway stations, which the July Monarchy had located at the limits of the old and completely saturated city centre. Industry and other polluting activities were moved even further out.[3] In this way, Haussmann's planning tried to find a solution for what seemed to be the main component of the mid-century urban crisis, the overdensification of the city centre.

The desire to make a break with earlier town-planning methods by modifying urban space in this way thus seems to have been essential, even if it was not the first time such an ambition had been expressed.[4] The novelty lay rather in the scope of the changes and the efficiency of the means used to carry them out. Even during the Second Empire, Haussmann's critics denounced the demolition of the old Paris and the destruction of its productive activities which had hitherto allowed 'the craftsman to live side by side with the financier'. In his *Comptes fantastiques d'Haussmann*, Jules Ferry deplored the disappearance of a city in which there were 'groups, neighbourhoods, quarters, traditions, and where the threat of expropriation did not permanently upset old relationships and the most cherished habits'.[5] He regretted the passing of a friendly, convivial city in which different populations lived together harmoniously and its replacement by a city of bourgeois residents living in modern apartment buildings. This analysis, put forward by the Empire's republican opponents, was widely shared by many other contemporaries, and subsequently by many historians. But is the picture it gives a true one? Were handicraft and industrial activities forced to leave the centre of Paris? Was the planned redistribution of functions operational over the whole, enlarged city by the end of the Second Empire? A different approach, using the methods of social history, brings answers to these questions which allow for more subtle shades in the depiction of urban reality.[6]

In this essay this line of inquiry will be pursued by studying the effects of urban renovation as it impinged on the economic activities of one of the neighbourhoods situated in the old artisanal and industrial heart of Paris, a neighbourhood which was radically transformed by Haussmannian urban renewal. The choice of this neighbourhood, the Arts et Métiers quarter situated to the north of the new (post-1860) third *arrondissement*, seemed pertinent.[7] At the beginning of the Second Empire, a direct tax inspector by the name of Maillard, responsible for drawing up some general statistics on the area, described its activities as follows: 'small-scale local production is primarily involved in fine gold jewellery and gold-plated jewels, comb-making, pasteboard articles, fancy goods, knickknacks and small turned or engraved bronze items'.[8]

These activities correspond with the trades collectively known as the 'Fabrique d'articles de Paris', the fabrication of Paris fancy goods. Another inquiry, published by the Chamber of Commerce in 1851, situated the quarter's activities within a broader geographical context: 'as for the fabrication of Paris articles, this is located to the north of the rue des Francs-Bourgeois and Saint-Merry, and extends over an area hemmed in by the rue Montorgueil and the rue Poissonnière to the west, and the place des Vosges to the east.'[9] These so-called Paris articles were goods of high quality, requiring not only technical skill but also first-rate workmanship. They included artificial flowers, toilet and vanity cases, boxes for photographers, combs, real and imitation jewellery and so on. They were expensive commodities, readily exported outside the capital. The local economic situation was, therefore, a relatively prosperous and dynamic one, which had not been too seriously affected by the mid-century crises.

On three occasions, then, this neighbourhood was the scene of urban renewal programmes, corresponding with the objectives of the first two campaigns mentioned above.[10] At the beginning of the Second Empire, the creation of the boulevard de Sébastopol destroyed a network of small back streets and passages situated between the rue Saint-Denis and the rue Saint-Martin (see Figure 11.2). The boulevard's new buildings went up very rapidly, but the whole thoroughfare was not completed by the date of its official inauguration on 5 April 1858. The decree which decided the creation of this boulevard, in the context of the first campaign, also envisaged the development of the properties along it. This development, carried out at the beginning of the 1860s, was the second phase of public works affecting the quarter, in particular the part of it situated to the west of the rue Saint-Martin. It involved the redevelopment of the rue Réaumur, the opening up of the beginning of the rue de Turbigo and a new, 20-metre-wide avenue opposite the Conservatoire. The latter avenue was never completed and the part which was begun was subsequently turned into a public square. The third and last phase of works in this neighbourhood was the most thorough-going, touching the quarter at its very centre. The rue de Turbigo was extended to the top of the rue du Temple, towards the place du Château d'Eau (today's place de la République), sweeping obliquely through the middle of the quarter and tearing its old fabric out from the south-west to the north-east. The rue Réaumur was lengthened towards the place du Temple, involving a straightening of the rue Phélipeaux. These operations, part of the second Haussmannian campaign, were carried out rapidly from 1865 on. Overall, then, the works affecting the quarter were of considerable scope, causing upheavals which lasted for 10 years.

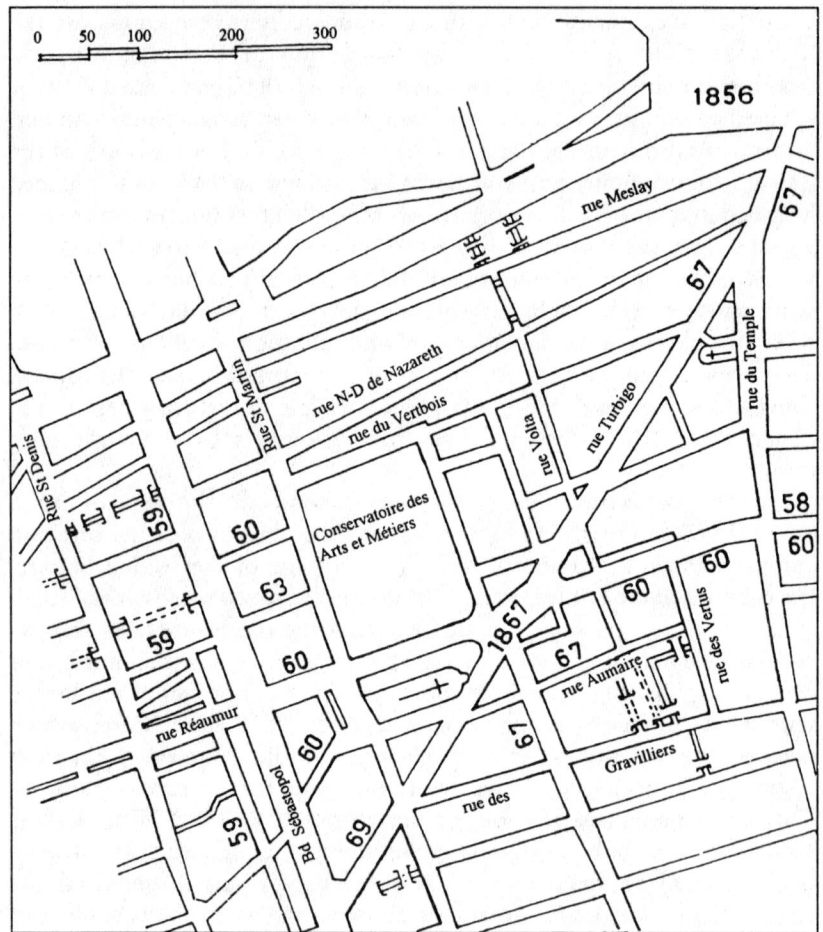

11.2 Major works in the Arts et Métiers quarter.
Source: A. Alphand, *Atlas des Travaux de Paris, 1789–1889*, Paris, 1889

How did this neighbourhood of handicraft and industrial producers react to its profound remodelling by Haussmannian town planning? Was local small-scale production particularly affected by the renovations? Did certain sectors benefit more than others? Were there significant alterations in the location of these activities?

Urban renovation and business

The opening up of the boulevard de Sébastopol coincided with a particularly prosperous period in the Parisian economy. The early years of the Second Empire witnessed an economic upswing which was already perceptible during the Second Republic. A measure of it is given by the continuing growth in the assessments for the *patente*. This was a municipal tax levied on all industrial and commercial activities. It dated back to the French Revolution as one of the new taxes created by the Constituent Assembly with a view to rationalizing the fiscal system as a whole. Along with the land tax, the personal and income assessment (*cote personnelle et mobilière*) and the tax on doors and windows, it was counted as one of the 'four ancient taxes'. Its evolution may serve to measure the intensity of local economic activity. The statistical report drawn up by the inspector Maillard details the returns of this tax at three-yearly intervals from 1846 to 1880.[11] It notes the firms active in the quarter, whether they belonged to local people or to owners from outside the neighbourhood,[12] and consequently allows us to follow the evolution of the economic situation.

New *patente* assessments appear from 1850 on, and, up to 1859, their number increased by 54.6 per cent, a rate of increase notably higher than that to be observed in other parts of Paris.[13] Some of the figures represent existing concerns, closed down by the crisis of the 1840s and then opening up again, but some of them are new establishments. The economic situation appears to have been favourable. The first campaign of public works, launched by the prefect Berger in 1852, the rise in demand resulting from the development of both home and export markets, money circulating more fluidly within the world of Parisian artisans: all these are factors which encouraged Parisian workers to set up in business. A law of 1858 modified the way the *patente* was levied, exempting artisans working on their own from payment of this tax. With the help of their families, these artisans could continue to satisfy the demands of commercial brokers, merchants or other, larger workshops without featuring in the *patente* tables. The number of assessments for this tax after 1858 is consequently an indicator which no longer accounts for the real extent of the quarter's activities. Even so, the overall increase in the number of assessments during this period is quite remarkable. In terms of the number of new establishments recorded, the peak years fell between 1850 and 1854, with an average of 250 new concerns every year. Subsequently this rate of growth slowed down somewhat, with an annual average of 88 new establishments until the exemption law of 1858. Growth still continued after 1858 and up to 1864 (+64 per cent in 1864, compared to 1850), and this despite a second modification to the *patente* law, enacted in 1862 and further exempting artisans employing only one worker.[14]

However, on the eve of the new works involved in the development of the rue de Turbigo and the rue Réaumur, the conditions for setting up in business became less favourable. The growth of the Parisian economy was adversely affected by rising rents and by the increasing cost of raw materials. The Paris Chamber of Commerce recognized that, even when the price of raw materials doubled, that of finished products only went up by 50 per cent.[15] The whole system for financing businesses was coming unstuck. Investors and wholesale merchants were less prepared to make loans as their own outlets narrowed. The artisan's state of dependence on a financial backer acted as an obstacle to recovery. Nonetheless, different sectors were more or less hard hit here, and the makers of Paris articles seemed to fare better than other trades in the capital, their goods rarely requiring expensive raw materials and being moved quite rapidly as 'fashionable' commodities.

The development of the rue de Turbigo and the rue Réaumur led to the disappearance of some 455 concerns assessed for the *patente*. In the year 1866 alone, 150 establishments were expropriated. The rue de Turbigo, cutting through the old centre of the quarter, and the widening of the rue Réaumur, put a large number of artisans, home-workers and tradesmen out of business, their homes expropriated. The worst year was 1866, but from 1867 on, new establishments were already opening.[16] The war of 1870, followed by the Commune, interrupted this recovery, and the figures of 1862 were only to be seen again in 1877. From this date, and up to the crisis of the 1880s, the quarter seems to have assimilated the public works of urban renewal and the strategic choices made within the sector of small-scale production.

Analysis of the *patente* assessment tables allows a closer observation of the changes in local patterns of production (see Table 11.1). Payment of the *patente* was divided into four main groups or tables, corresponding to different types of activity. The eight categories of Table A covered small and medium-sized shops and artisans, Table B large businesses, Table C industrial concerns and Table D the liberal professions. The latter were taxed more systematically during the Second Empire in order to compensate for the loss of revenue resulting from the exemption of the smaller *patente*-payers in 1858, 1862 and 1868. The different payers of the tax were distributed over the four tables and their respective categories according to the nature of their activities and the rental value of their professional premises and their dwellings.[17] Only the first three tables are of interest for our present argument. All three show the expansion in the years 1850 to 1864. The first categories of Table A and all those of Table B, however, show particularly strong growth. These are primarily large-scale commercial ventures, and their development accompanied the general growth of the number of taxpayers in the categories covering handicraft production and small and medium-sized shops. There were still close links between the

Table 11.1 Tables and categories of the *patente* in the Arts et Métiers quarter, 1852-79

	1852	1855	1858	1861	1864	1867	1870	1873	1876	1879
1e	12	10	20	27	28	37	33	30	32	19
2e	32	27	28	46	50	47	102	107	103	98
3e	71	80	95	136	143	134	135	165	168	154
4e	183	209	230	292	296	261	347	360	362	340
5e	363	432	464	544	546	509	477	525	522	493
6e	657	861	879	903	913	865	722	748	757	709
7e	832	970	1085	1019	1027	836	786	746	826	899
8e	248	276	323	312	318	272	223	196	204	229
Total A	2398	2865	3124	3279	3321	2961	2825	2877	2974	2941
Total B	8	10	10	19	26	29	20	26	16	13
Total C	13	18	13	28	20	16	21	25	24	-

Note: See text for explanation of A, B and C.

Sources: Archives de Paris, D3P4, Statistical Register; D3P2, *Etat du montant des rôles*.

independent producer or artisan and the investor or middleman who purchased from him. The number of industrial establishments featuring in Table C also grew during this period: in 1861 there were 28 industrial establishments registered in the quarter. On the whole, the proportional relationship between industrial and artisanal activities remained stable: in 1852 there were 63.8 artisanal concerns for every industrial establishment, and in 1864 the ratio stood at 60.1 to one. The increase in average rental values, also recorded for the calculation of the tax, was lower for the industrial establishments of Table C than for the artisanal ones of Table A. No major industrial concerns left the quarter during these years and, indeed, several new industries moved into the neighbourhood from nearby *arrondissements*: five metallurgical firms, three in textiles and leather, one in the chemicals sector, two in the furniture trade and one in public works. The quarter could still boast an industrial potential. The years of expansion, 1852-64, therefore correspond to a growth in large-scale commercial firms, nourished by handicraft and industrial activities.

What happened after 1864? *Patente*-payers diminished in all the categories of Table A, by 10 to 20 per cent amongst the upper categories and by as much as 30 per cent for categories 7 and 8. The war and the Commune accentuated the disparities between the top and the bottom of Table A. The recovery, which is visible from 1867, was due to the progress in wholesale activities and wholesaling in small quantities (*demi-gros*). The complementary relationships between handicraft production and large-scale commerce were modified as wholesale merchants, brokers and bankers moved in. The traditional buyers of the Paris articles, investors who kept a close watch over 'their' workers, supplying them with raw materials and even controlling to a certain extent the fabrication processes, now gave way to merchants, middlemen who placed their orders with the artisans and served as intermediaries between the producers and the retailers, or even, in some cases, between these producers and the new Parisian department stores. Gradually, the traditional pattern of Parisian production, the *fabrique*, to which the renewed demand of the Second Empire had given a new lease of life, was replaced by a sort of 'sweating system' of industrial work put out.[18] The changeover is a slow one, however, and the same taxpayers who are denominated merchant-producers (*marchands-fabricants*) subsequently appear as wholesale merchants (*marchands de gros et demi-gros*), often operating from the same premises. This renewed form for the way goods were ordered and work carried out 'in crumbs',[19] parcelled out into tiny batches, created the conditions for economic recovery. After 1876, this recovery largely concerned the establishments in the lower categories of Table A, the small workshops and shops.

The industrial concerns featured in Table C were less dramatically affected by the public works carried out in the quarter. The addresses of these concerns are known to us by the 'industrial notebooks' (*calepins industriels*), which describe the premises and give an estimate of the value of the works and of its machinery.[20] Most of them were located to the north of the quarter or in the old streets of its centre. This position meant that many of them escaped the expropriations resulting from Haussmann's works. If their number declined - in 1867 there were 29 industrial establishments, in 1870, 21 - this was on account of one of the indirect effects of the urban works: space for industrial activity was becoming short and rents were becoming too high. The largest works - copper foundries manufacturing copper light-fittings or 'objects of art and ornamentation', and rolling mills - left the neighbourhood. But the proportion of industrial concerns in relation to those of artisans rose: there were 48.2 of the latter for each industrial enterprise in 1870 and 42.8 in 1878. This suggests that new forms of cooperation were evolving between the two sectors. Unlike handicraft activities which had to stay in the quarter in order to survive, industry left the quarter and did not come back. This was not only a geographical displacement; it was also a change in the nature of the quarter's industrial vocation. The new establishments, and those which remained in the neighbourhood, were of smaller size than previously. After a rise, due to the pressure on industrial space during the Second Empire and immediately following the war and the Commune, the average rental value of these establishments tended to decline between 1873 and 1879.

Local activities do not appear to have suffered too much from the urban renovation of the Second Empire. With the exception of the short period of expropriations, commerce and handicraft production benefited from the situation to pursue their development. The changes which industrial activities were forced to carry out were more thoroughgoing, but in both sectors what is most striking is the capacity of both to adapt to a new situation. What factors made this adaptation possible?

Changes in the Parisian *fabrique*

We have seen that the patterns of production known as the *fabrique* of Paris represented the main activity of this neighbourhood. The period of Haussmannian urban renovation was marked by significant changes in these patterns and in the sorts of goods produced. The quarter's occupational milieu was able to survive, maintaining its activities and its specificity, by applying a strategy of diversification.

The largest number of concerns paying the *patente* was grouped in categories 6 and 7 of Table A, and in what follows we shall concentrate on these categories. Information about them comes primarily from the statistical report drawn up by the tax inspector Maillard.[21] This report gives a complete list of professions and specifies the number of fixed assessments, corresponding to the number of enterprises in the quarter. Within the different professions listed, this information allows us to distinguish between handicraft production and commercial activities, although of course the difference between making goods and selling them is not always clear-cut: both activities were often carried out in the same shop. For our analysis we have, therefore, tried to identify activities which are of a purely productive nature, pursued in specific premises.

As is to be expected, the period of economic growth between 1852 and 1864 witnessed a marked increase in the number of these artisanal concerns, rising by 54 per cent in category 6 and by 36 per cent in category 7. These percentages do not take into account the exemption from payment of the tax accorded to the smallest craftsmen by the legislative changes of 1858 and 1862. In both categories the number of commercial ventures is scarcely larger in 1864 than in 1852. What growth there is in the shopkeeping sector is most noticeable in category 6, suggesting bigger shops, moving up the scale towards the middling categories. Newer shops seem to fare better than the older-established ones.

In 1873, after the war and the Commune, this upgrading of commerce continued. At this date, there were only half as many small shops as in 1852 and the development of such small shops only began again in 1876. The larger shops of category 6 did not recover the lost ground, a real change of scale having been imposed by the rise in rents and by the opening up of the new thoroughfares. As for handicraft activities, although their number is not as high as in 1864, they are still numerous after 1870.

The period of the strongest growth in the number of concerns paying the *patente*, before the first exemption law of 1858, is also the period during which the diversity of trades is most striking. The year 1858 coincides with the greatest variety and innovation in the world of artisans, with a total of 98 distinct trades listed in category 6 and 91 in category 7. In the latter category, the Parisian *fabrique* represented nearly 40 per cent of the professions, accounting for 50 per cent in category 6. The *fabrique* comprises 43 different trades: the manufacture of artificial flowers and leaf-work, of knick-knacks, umbrellas, toilet and vanity cases, combs, etc. Along with these fancy goods, precision metallurgy may also be included, turning out rims for spectacles, rings and so forth. The increase in the number of artisans also bears witness to the introduction of new trades. The raw materials given to the artisans by the middlemen who placed their orders with them were worked into products which relied on the manual skill – the 'tricks of the trade' – of a specialist craftsman with real know-how. The type of work done evolved during these years, which were those of the heyday of the Second Empire, the *fête impériale* which generated new demands and new pressures. This evolution involved the introduction of new techniques and new production processes, technical innovation in the *fabrique* affecting dyeing, fabric-finishing, electroplating and veneering.

One answer to the problems jointly posed by rising demand and by the introduction of new techniques was the extreme division of work tasks. In order to produce greater quantities without having to replace his tools too frequently, the artisan parcelled out the production process into a sequence of simple operations.[22] Jeanne Gaillard has looked at this specialization in the world of artificial flower production.[23] The 'roser', who specialized in the fabrication of artificial roses, only assembled the flower from pieces which had already undergone several preliminary operations, and this 'roser' himself only produced certain sorts of artificial roses! In his presentation of the Chamber of Commerce inquiry of 1861, Augustin Cochin describes this division of work in a sector similar to that of the *fabrique* – the production of shawls. Quoting the report prepared by Parisian workers for the London exhibition, he regrets this extreme division of workshop activities and the negative effects it could have on professional training:

> the shawl designer divides up the work between eight different workers: the worker who composes the first design, the one who does the rough drawings, the one who then enlarges them on card, the one who picks out the main lines of the design, the one who chequers the outlines, the one who draws in the details, the one who transfers these details to the card and the one who fills them in. The apprentice becomes highly skilled, but is familiar only with one eighth of his trade.[24]

This division of work tasks was to be seen not only at the level of the individual workshop but also throughout the neighbourhood. The same object went from hand to hand and from workshop to workshop, to be completed and receive its final label as a 'Paris article'. Taking advantage of the economic upswing of the early years of the Empire, many workers set up on their own in business, exploiting the special skills they had picked up in another workshop. The consequent hyper-specialization in tasks explains the tremendous diversity of trades observed by the inspectors of the direct tax administration, and gives evidence of the ease with which the *fabrique* of Paris articles could reorganize its production system to meet rising demand. Work carried out in a myriad of tiny and highly specialized workshops allowed for remarkable flexibility in the face of the fluctuations of the Paris market, then booming.

The spatial and functional diversification of the activities of the *fabrique* is also to be seen in new forms of cooperation with the other sectors of local activity. After the *fabrique*, the second place in the neighbourhood was occupied by the metalworking trades. In 1852, these represented 32.6 per cent of the *patente*-payers and accounted for 19 different trades: workshops specialized in the production of bronze objects, of nails and screws, alongside turners, wire-drawers and tinsmiths. Much of this metalwork went into the production of Paris articles, in the form of frames for bags or for umbrellas, mountings for rings and so forth. The third sector active in the quarter, clothes-making and the fashion trade, was also close to the fashion-conscious world of the Paris article.

From 1858, and more particularly after the works of urban renovation, significant changes took place in the way the handicraft trades of the quarter were distributed. The laws of 1858 and 1862 had some influence on these changes, but were not the only factor since the same changes are to be seen in categories 6 and 7 of those not exempted by these laws. These changes are characterized by a steady decrease in the number of *patente*-payers involved in metalworking and their gradual replacement by trades associated with clothes-making and fashion. This modification in the activities occupying the second place in the quarter, after the fabrication of Paris articles, is partly the result of the development of this latter, predominant sector. There is no reason why the producers of metal artefacts should have been more adversely affected by expropriations than others. The reason for their relative decline is to be found rather in the competition now coming from more dynamic sectors. The only metalworking activities that stayed in the quarter were those directly involved in producing goods required by these other sectors: mechanics specialized in sewing machines or steam engines, or polishers and manufacturers of settings for imitation jewellery.

Within the world of the production of Paris articles, there was a new distribution of trades. After having first encouraged diversification, associated with the development of the quarter's other activities, the *fabrique* now reduced the range of goods it produced. The more dynamic trades witnessed a process of stabilization between 1858 and 1864, a process which the public works campaigns tended to reinforce. The recovery of 1867 seen in the trades of the Parisian *fabrique* is mainly the result of the development of its two most prosperous activities, the making of artificial flowers and leaf-work, and the production of imitation jewellery. Next came the makers of toilet and vanity cases (now branching out into boxes for cameras), followed by umbrella-makers. These changes did not then affect the contents of the trade but denoted a clear choice for greater specialization in the goods produced. Production was still carried out in small and medium-sized workshops. The traditional patterns of production thus survived, better adapted to market conditions. The changes did not impinge so much on the scale or the structure of the enterprises as on their forms of cooperation with the agents and investors who ordered their merchandise.

The mutations in handicraft production were confirmed by the choices to be seen in local industrial activities. The industries of the neighbourhood feature not only in Table C of the *patente* assessment but also in its 'industrial notebooks' (*calepins*).[25] At the beginning of the Second Empire, the metal-based industries, with six copper foundries, two rolling mills and a filemaking works, held first place in the neighbourhood. The second place was occupied by industries associated with garment manufacture and *passementerie*. There was a feather-dyeing works, which remained in activity throughout the period, and another works producing lace and woollen braiding. Another firm built looms. In 1855 a new establishment was opened in the rue Aumaire, specializing in the making of ivory combs and marking the beginning of the industrialization of the fabrication of 'Paris articles'. The industrial production of these articles accounted for almost a quarter of the firms listed in Table C of the *patente* assessment in 1861, at the expense of the older metalworking concerns. As in handicraft production, the early 1860s saw considerable diversification in these industrially produced articles. New works opened up, many of them to disappear from the registers by 1864. There were works for the production of glue, for the making of printers' ink and for the manufacture of umbrella frames – in all 17 different types of product, all of them, with the exception of the copper foundries, supplying the small artisan producers. None of these local industries was established on a large scale. The largest in the neighbourhood, according to the industrial notebooks, was situated at 5–9 rue Sainte-Elizabeth, a factory producing lighting materials and plumbing accessories and employing 100 workers, either on the premises or outside. Its equipment was relatively simple: 50 vices, 22 ordinary mills, 12 forges, etc. This was production concentrated in one place, rather than a modern industrial enterprise. The other principal industrial concern of the quarter was a rolling mill with a steam engine, but, here again, its modest size leaves it in the realm of the large workshop.

The reconversion of the quarter's industrial activities seems to have been accelerated by the public works campaigns of the Second Empire, and the pattern of new activities replacing old ones continued. In 1876 there were no more metalworking firms in the quarter. Replying to the question 'Are small industries surviving or are they disappearing?', a local police inspector was probably thinking of these metalworking concerns when he wrote in a report on the quarter's industry, 'some are surviving but others are declining day by day ... because of the competition from large-scale industry'.[26] At first, and practically at the same time as the urban renovation work, the change affected the mechanization of the production of Paris articles. The new establishments – works for making fans, tortoiseshell combs, horsewhips, pasteboard articles – were little more than large-scale workshops, combining work on the premises and work put out. The major difference lay in the new equipment of these works and in their use of steam power. After 1870, the new industrial establishments – six in 1870 and 13 in 1873 – were connected rather with the garment and fashion trades. The industrial notebooks of 1858 and 1867 both describe a fabric-finishing works which occupied a whole building at 86 rue Notre-Dame de Nazareth. Under new management, it became one of the most successful and concentrated enterprises of the quarter. The 'market value' of its production equipment, used for calculating the *patente*, rose from 9900 francs in 1858 to 30 000 francs in 1867. In the rue Saint Martin, the *passementerie* industry was transformed by the introduction of steam power. At certain stages in the production process, concentration thus became a necessity. The work put out to domestic weavers in the street was of another sort, and the industrial notebooks ceased to mention the existence of these enterprises which thus associated the small producers of the neighbourhood with those of other parts of Paris or in the provinces. Local production was also given new life by the demand coming from wholesale commerce and the mixed retail and wholesale establishments of the boulevard de Sébastopol. Buttonmakers, threadmakers, embroiderers and *passementerie*-makers all worked for orders placed by these merchants on the boulevard.

At the end of the renovation of the quarter, then, industrial activities were concentrated around two main poles, the Parisian *fabrique* and production associated with making clothes.[27] Between these two poles, there were also new, specialized activities. As in handicraft production, the choice of the *fabrique* favoured jewellery production, both real and imitation. In terms of the value of their equipment and the number of workers employed, the largest establishments of the quarter were the gold-beaters, one of which was located in the rue des Gravilliers.

The feather-dyeing works situated in the rue du Vertbois is a special case, both in terms of its long survival and its continuation without modernization of its equipment or its working methods. The equipment, detailed in the industrial notebooks, comprised copper boilers, dryers and basins. The works employed five, then seven workers. Its continuing existence constitutes a striking example of the way the industrial and handicraft sectors within the *fabrique* could complement each other. Producing only on a very modest scale, it nonetheless managed to satisfy the rising demands of the fashion trade. The work was arduous and delicate, but unlikely to come into competition from larger concerns. It represents one of these Parisian industries that remained inside the capital, producing semi-finished goods designed to meet a specific demand.

In this way, local production underwent partial renewal, developing the more successful sectors in its speciality. The fabrication of Paris articles adapted itself, enabling small workshops to survive, their flexibility constituting a considerable advantage, and also allowing associated industries to prosper. The works of urban renovation also played a part, by changing the conditions in which these activities took their place in the space of the city, modifications which we shall now examine.

Urban space and professional space

In the introduction to his statistical register, the inspector Maillard made a distinction between the streets of the old centre of the quarter, lined by dwellings described as 'workers' houses', and the periphery, with its more prestigious streets lined by finer residential blocks - streets such as the rue Meslay, which played a dynamic role in the commercial animation of the quarter, along with the rue Saint Martin and the rue du Temple. This differentiation, which underlay his plea for sanitary improvements in the neighbourhood, is also to be found in the analysis of the localization of different types of trade. The mid-century Parisian administration carried out a systematic survey of each property with a view to completing the capital's cadastral survey, documents often erroneously termed the 'notebooks for the revision of the cadaster'. These documents give a description of the property and the name and profession of its owner, detailing its rental value which was used for calculating the level of *patente* to be paid and the category under which the taxpayer should be assessed.[28] We have examined these notebooks for a dozen streets in the neighbourhood for the years 1852 and 1876. In 1852, the number of concerns assessed for the *patente* for every 100 metres of street works out at 55.2. The streets with the greatest density of these taxpayers were those in which handicraft and industrial activities were also the most numerous - rue Aumaire, rue Volta, passage de la Marmitte, etc. - the central streets of the quarter. The streets with fewer of these *patente*-payers are the 'shopping' streets.

This observation must be qualified, however, by comparing the different rental values for commercial premises and artisans' premises in ten of the quarter's streets. The difference between the streets is still considerable, but there is also a difference between rents paid in artisanal premises and in commercial ones. For all the categories of Table A, the coefficient of variation in rents paid by artisans is only 0.15, no matter what the street, whereas the coefficient is more than 2.0 for shop premises. Thus the most notable difference between the streets relates to commercial activities and corresponds to the difference between a small shop and a warehouse. In other words, there is less difference between a workshop operating at the bottom of a courtyard behind an elegant apartment block in the rue Meslay and another workshop active in a single apartment in the rue Aumaire than there is between two shops in these same streets.

The detailed localization of the quarter's different activities of production may be summarized in other terms. In 1852 the production of Paris articles is spread all over the neighbourhood, clearly dominating the 'industrious' streets of its centre. There is an enormous variety in the types of activity and types of goods produced: the rue Volta and the rue Aumaire housed between 22 and 25 different types of activity belonging in the general realm of the *fabrique* of Paris. Only jewellery seems to have a certain geographical specificity, being the speciality of three or four particular streets. Between 15 and 18 per cent of the *patente*-payers are involved in the real or imitation jewellery business in the rue Phélipeaux, the rue Réaumur and also the rue Saint Martin and the rue Meslay. In these streets, average rental values are higher and the types of fabrication less numerous. The secondary activities of these streets are also interesting to note. In the central streets, the work of the *fabrique* is closely associated with metalworking; in the rue Meslay, the association is with garment production and the fashion trade; in the rue Saint Martin the association is with wood and also with all the other specialities of the *fabrique*. This last street served as a kind of 'shop window' for all the local activities. What happens to this situation after the urban renovation of the Second Empire?

In 1876, and despite the three laws exempting certain artisans from paying the *patente*, the density of activity had increased even further. The average is now 64.4 *patente*-payers for every 100 metres of street. At the same time the population of the quarter has declined (from 30 101 in 1851 to 26 579 in 1876), which means that the links between the population and its productive activities are even more intimate. Within the quarter as a whole, the comparison with the situation in 1852 is rendered difficult by the existence of the new streets and the rue Réaumur which had undergone enlargement and development. The disparity between the old streets, even the most elegant amongst them, and the new ones is, of course, striking. More striking, however, is the continuity of solidarity between different streets after the imperial works of urban renovation. Thus, not counting the newly created streets, there is a still a clear grouping of streets with high average rental values, including the rue Meslay, the rue Saint Martin and the rue Réaumur, situated on the 'outskirts' of the quarter, whilst another group is comprised of the quarter's central streets. This continuity may be explained in part by the continuity of the activities, carried out in these streets. The old, elegant streets have adapted themselves, accommodating new artisanal activities and those in particular – like the jewellery trade – which were most prosperous, stimulated by the new shops and workshops of the new thoroughfares. At number 22 rue Réaumur, for example, we find all the trades involved in imitation jewellery, from the fabrication of false stones, to the setter, the stone-merchant and the specialist retailer. Their position is a reflection of the choice made in favour of the most dynamic sector of the *fabrique*. In the old streets, however, the situation is different, these streets seeming to serve as a refuge for all the other products classified as Paris articles. There is a bit of everything here: purses, wallets, umbrellas, 'accessories for umbrella handles', knickknacks and cheap baubles, these goods no longer so profitable during the 1870s. It is also in these old streets that we find the last representatives of the metalworking trades: engravers and embossers and makers of objects in polished steel. Most of their merchandise is oriented towards the demand for semi-finished products used by the imitation jewellery trade. Thus the urban changes have perpetuated the forms of complementarity which existed between the different streets at the beginning of the Empire, even if the contents of this complementarity have been significantly altered.

Between 1854 and 1867, the Arts et Métiers quarter was the scene of extensive works of urban renovation. It is difficult to imagine what the conditions of everyday life were like during these operations, amidst all the building sites and piles of rubble. For a period of at least 13 years, construction work followed demolition work in the opening up of new thoroughfares. Rehabilitation operations carried out in the existing streets did not simplify the situation. Yet, despite all these major works, the sector of the Parisian *fabrique* witnessed considerable renewal, and to a large extent the local population was able to remain in the neighbourhood or to go no further than the neighbouring quarters if forced to leave. Local dynamism was, therefore, particularly remarkable, even though this dynamism does not explain everything. The correspondence between urban renovation and the changes in the local economy is striking and may itself be seen as a source of this dynamism. The public works, in this neighbourhood, contributed to the renewal of patterns of production. Expropriations played an obvious part, imposing reorganization or changes of address necessitated by higher rents and by the competition for professional premises still available. Above all, however, the creation of new thoroughfares enabled this quarter to be opened up and integrated into an infinitely larger Parisian market. The dynamic qualities of the local business milieux are consequently part and parcel of the changes wrought on urban space. This interpretation is a far cry from the traditional vision of the effects of Haussmannization. A popular and industrious urban milieu not only survived these operations but even profited from them. Can this same phenomenon be observed elsewhere in Paris? It is not clear, and the continuation of research at a local level becomes all the more valuable here. From observations in the Saint-Victor quarter, on the left bank, it seems that the effects of Haussmannian urban renovation could also be a real 'dispossession'.[29] This quarter was not a centre of handicraft activities and, if it served at first as a kind of refuge for these, it subsequently became a zone of bourgeois or semi-bourgeois residence. As a counter-example, it serves to underline the interest of the correspondence that has been noted between economic dynamism and the 'success' of the renovation in the Arts et Métiers quarter, illustrating the particular capacities for resistance on the part of a coherent and active professional milieu.

Notes

1. See Figure 11.1, below.
2. There are many studies of the urban renewal of Second Empire Paris. See, in particular, Jeanne Gaillard, *Paris, la ville, 1850-1870*, Paris: Champion (1977), pp. 6-66, and the articles by Marcel Roncayolo and Louis Bergeron in Louis Bergeron (ed.), *Paris, genèse d'un paysage*, Paris: Picard (1989), pp. 218-42.
3. From 1867, the city limits for the *octroi*, the toll on goods entering the capital, were extended beyond the first suburbs, forcing Parisian industrialists to pay this municipal tax on their raw materials. See Gaillard, *Paris, la ville*, p. 60.
4. See Guillaume de Bertier de Sauvigny, *La Restauration, 1815-1830*, Paris: Association pour la publication d'une histoire de Paris, Hachette; and Philippe Vigier, *Paris pendant la Monarchie de Juillet*, Paris: Association pour la publication d'une histoire de Paris, Hachette (1991), pp. 177-220. See also the catalogue edited by Jean Des Cars and Pierre Pinon, *Paris-Haussmann*, Paris: Edition du Pavillon de l'Arsenal et Picard (1991), pp. 21-62.
5. Jules Ferry, *Les Comptes fantastiques d'Haussmann*, Paris: (1868).
6. See the important work done by Adeline Daumard, in particular her *Maisons de Paris et propriétaires parisiens au XIXe siècle, 1809-1880*, Paris: Édition Cujas (1965). On a more detailed level, see Gérard Jacquemet, *Belleville au XIXe siècle, du faubourg à la ville*, Paris: Éditions de l'École des Hautes Études en Sciences Sociales (1984), and also Gaillard, *Paris, la ville*.
7. The administrative limits of this quarter were only very slightly altered when the *arrondissements* were modified and renumbered in 1860, making comparisons before and after this date possible.
8. *Archives de Paris*, D3P4, statistical register for the Saint-Martin des Champs quarter, before 1860, third part.
9. Chambre de Commerce de Paris, *Enquête sur l'industrie*, Paris (1851).
10. See Figure 11.2.
11. The official returns distinguish between 'fixed rates' and proportional rates, the former corresponding to businesses installed in the quarter and the latter to these same businesses, along with commercial and industrial premises belonging to taxpayers whose fixed rates are located elsewhere, outside the quarter. The figures, therefore, cover two separate realities, the first indicating the number of firms actually located in the quarter, the second the total number of firms trading in the quarter. Under the heading 'article', the register also gives the total number of *patente* returns, based on the fixed and proportional rates, and other associated

rates. The same total is also given by another source, the annual *État du montant des rôles*.
12. *Archives de Paris*, D3P4, statistical register for the Saint-Martin des Champs quarter before and after 1860, and D3P2, annual *État du montant des rôles*.
13. Between 1850 and 1858, the rise in the number of concerns assessed for the *patente* in four other quarters of Paris are as follows: École de Médecine - 27.7 per cent; Monnaie - 43.7 per cent; Mail - 45.2 per cent; Montmartre - 50.9 per cent. The first two are situated in today's sixth *arrondissement* on the left bank, the two others in the second *arrondissement* on the right bank. See Florence Bourillon, 'Grands travaux et dynamisme urbain: Paris sous le Second Empire', in A. Faure, A. Plessis and J.C. Farcy (eds), *La terre et la cité, mélanges offerts à Philippe Vigier*, Paris: Créaphis (1994).
14. See Jeanne Gaillard, 'Les intentions d'une politique fiscale: la patente en France au XIXe siècle', in *Bulletin du Centre d'Histoire de la France contemporaine*, 7, 1986, pp. 28-9; and Alain Faure, 'Note sur la petite entreprise en France au XIXe siècle. Représentation d'État et réalités', in *Entreprises et entrepreneurs, XIXe-XXe siècles*, Congrès de l'Association des Historiens économistes, March 1980, Paris: Presses de l'Université de Paris-Sorbonne (1983), pp. 203-5.
15. See Gaillard, *Paris, la ville*, p. 390.
16. The effects of the new *patente* law of 1868, exempting from payment of the tax artisans employing only one apprentice, are not visible because of the declining returns in the years just before this law.
17. The summary tables indicate the rental values used to calculate the rate at which the tax was levied. The inclusion of firms in Table A depended on the nature of the activity carried out - bankers and small-scale commercial brokers for example figure among the higher categories of Table A - and on the declared rental values. In the lower categories (7 and 8) the average rental value was much lower than in the upper categories of the same table. See Faure, 'Note sur la petite entreprise', p. 205.
18. Jean-Michel Gourden, *Le peuple des ateliers, les artisans du XIXe siècle*, Paris: Créaphis (1992), p. 53ff. See also the special number (108) of the review *Le Mouvement social*, 'L'atelier et la boutique', July-September 1979, and Geoffrey Crossick and Heinz-Gerhard Haupt (eds), *Shopkeepers and Master Artisans in Nineteenth-century Europe*, London: Methuen (1984).
19. Alain Faure, 'Petit atelier et modernisme économique: la production en miettes au XIXe siècle', in *Histoire, économie et société*, 4, 1986, p. 533.
20. These industrial notebooks were drawn up on two occasions for the Arts et Métiers quarter, between 1858 and 1861 and between 1868 and 1869. This source also gives the name of the owner or manager, the number of workers employed (specifying whether they worked on the premises or outside), the position of the machinery, etc. Rental values are also noted, since these are used in calculating the amount of *patente* to be paid. They comprise the sum paid in rent for the professional premises, the rent paid for the taxpayer's dwelling, along with the value of the machinery. These industrial notebooks are consequently a mine of valuable information, but unfortunately have not survived for the whole of Paris. The notebooks for the Arts et Métiers quarter are in the *Archives de Paris*, D2P4 16, 1858-61 and 1868.

21. *Archives de Paris*, D3P4, statistical register.
22. Faure, 'Petit atelier et modernisme économique', p. 538; Alain Dewerpe, *Le monde du travail en France 1800-1950*, Paris: A. Colin (1989), p. 78.
23. Gaillard, *Paris, la ville*, p. 438.
24. Augustin Cochin, *Paris, sa population, son industrie*, Paris (1864).
25. *Archives de Paris*, D3P4, statistical register, and D2P4 16, *Calepins industriels*.
26. *Archives of the Prefecture of Police*, Paris, BA 400, reports by police inspectors on workers, for the parliamentary inquiry of 1872.
27. Many of the new shops which opened up along the rebuilt rue Réaumur, the rue de Turbigo and the boulevard de Sébastopol were associated with the garment industry: hatters, drapers, fabric shops and clothes shops. See *Archives de Paris*, D1P4, notebooks for the cadastral revision for 1862 (boulevard de Sébastopol) and 1876 (boulevard de Sébastopol, rue de Turbigo and rue Réaumur).
28. *Archives de Paris*, D1P4, notebooks for the cadastral revision for 1852 and 1876 (rue Aumaire, rue Henry, rue Meslay, rue Phélipeaux, rue Réaumur, rue Saint Martin, rue du Vertbois, rue Volta, passage de la Marmitte, boulevard de Sébastopol and rue de Turbigo).
29. See Florence Bourillon, 'La rénovation du quartier Saint-Victor sous le Second Empire', in Nanterre-Paris X, *Recherches contemporaines*, 2, 1994, pp. 79-112.

CHAPTER TWELVE

Artisans and the labour market in Dutch provincial capitals around 1900

Pim Kooij

Industrialization in The Netherlands started relatively late. In his classic book about Dutch industrialization J.A. de Jonge situates the 'take-off' at around 1895.[1] Of course this does not mean that before that time there was no modern industry at all. In an article, de Jonge himself has pointed out some successful early initiatives in the Dutch countryside: the cotton industry which started around 1830 centred on the towns of Enschede and Hengelo; the woollen industry in the south centred on Tilburg; the potato-flour industry (1840) and the manufacturing of strawboard (1870) were created in the newly cultivated peat region in the province of Groningen in the north.[2]

Most modern industry, however, was established in the cities. J.L. van Zanden has in a more recent publication described some early industrial activities in Amsterdam.[3] The steam engine was, for instance, introduced from 1820 onwards in shipbuilding and machine manufacture, rice-polishing, bread factories (not very successfully), and the production of gas. Most of these plants were connected to the Amsterdam staple market, which in the 1850s lost its importance. As a result large new factories concentrated on the internal market: steam breweries, cigar factories, printing companies, bread factories (again), and building, although in this case without steam. In the 1870s the diamond industry became important. For a time these branches mitigated the consequences of the Great Depression for Amsterdam, but in the 1880s the Amsterdam economy was hit by the declining business cycle in such a way that recovery was only possible through structural changes.

These structural changes were characterized by the emergence of a dual economy, consisting of a modern industrial sector and a traditional crafts sector. Modern capital-intensive industries were sugar-refining, beer-brewing, and the manufacture of ships and machinery. To a certain degree they also included factories already in existence which had reorganized themselves. Access to the capital market was relatively easy and relatively high wages were paid. On the other side stood the traditional craft sectors, where competition was high and the size of firm rather small. The gas engine and electro-motor had allowed some mechanization to take place, but most of the activities were still performed by hand. Wages in this sector were much lower as a result of the large supply of labour which was increased by immigration.

Furthermore, Ad Knotter has established that this dual economic structure also included a dual labour market.[4] This non-integrated labour market also consisted of a traditional sector where working conditions were bad and wages were low, and which functioned as a reservoir for the modern sector. Youths were trained there and sometimes obtained the opportunity to enter the modern growth sector. Immigrants constantly filled this reservoir, because they did not have many chances to penetrate the modern sector, where the indigenous workforce kept the jobs for themselves. At first most migrants, mainly originating from the North Holland countryside and the province of Friesland in the north of The Netherlands, found work in the construction industry, mainly in speculative jerry-building. When this branch collapsed in 1883 immigration nevertheless continued, due to the agrarian depression which diminished opportunities in the villages. At that time the docks offered some additional employment, thanks to the opening of the Noordzeekanaal in 1876, but unemployment remained at a high level. Some relief was offered by the new clothing industry, also a modern branch but not so highly mechanized, where the wives and daughters of unemployed labourers could earn some kind of a living. At the same time, however, this caused unemployment among independent tailors. Only a few of them ever got the chance to become cutters in the clothing industry. Around 1895, when the agrarian depression was over, the character of immigration changed somewhat. More immigrants began to come in from other cities and towns. A number of them were skilled artisans, who could fill the gaps caused by the Amsterdam-born people who had crossed over into the modern sector, but there was not enough room for all of them.

So far Amsterdam is the only city where the labour market in the second half of the nineteenth century has received detailed attention. For Rotterdam, the second city of The Netherlands, Henk van Dijk also found a labour reservoir during the agrarian depression, as well as growing demand in the docks, but the industrial labour market was not a part of his analysis.[5] The question therefore arises of whether this dual economic structure combined with a dual labour market was also to be found in other Dutch cities.

The provincial capitals are especially interesting in this respect. During the French occupation (1795–1813), the Dutch Republic, which consisted of a rather loose federation of semi-autonomous provinces, was transformed into an integrated kingdom. The national government saw the creation of an integrated transport network as one means of promoting its idea of unity. Highways were constructed to link the provincial capitals, while the existing canal system was improved by digging several new canals for long-distance traffic. This created opportunities for local industries, which until then had primarily produced for regional economies, to operate in a national market. Two factors in particular proved to be a stimulus for this kind of specialization. The first was the loss of Belgium in 1830. This southern section of the kingdom of The Netherlands was destined to become the industrial part, with the consequence that the north now had to industrialize itself. In fact the industrialization in rural areas received a marked stimulus from this separation.

The second stimulus was railway construction. Railways had been initially confined to the western part of the Netherlands, with private investment in other parts of the country failing. As a result, the government started to finance railway construction in the 1860s. By 1870 the north, east and south were connected to the Dutch railway system, and the subsequent construction of secondary lines and tramways completed the integration of the infrastructure.[6] I have described elsewhere the importance of railways in the creation of an integrated urban network system in The Netherlands.[7] Since industrialization in The Netherlands, with some exceptions, took place in towns and cities, the completion of the railway network, which coincided with the end of the agrarian depression, increased the possibilities for urban economic specialization. This was especially true for the capitals of the so-called 'land provinces' of Friesland, Groningen, Drenthe, Overijssel, Gelderland, North-Brabant, and Limburg. Their respective capitals – Leeuwarden, Groningen, Assen, Zwolle, Arnhem, 's-Hertogenbosch and Maastricht – combined their functions within the urban network system with their position as a regional capital. It is interesting to ask whether the new opportunities for specialization in the regional capitals also created a dual economic structure with modern industries and other companies oriented on the (inter)national market on the one hand, and a more traditional regional sector on the other hand. The second question must of course be whether this structure was accompanied by a dual labour market.

The economic structure of Dutch regional capitals around 1900

In 1900 the number of inhabitants of the provincial capitals mentioned above varied between 30 500 in 's-Hertogenbosch and 66 500 in

Groningen. Only Assen, the capital of the province of Drenthe, was much smaller with a population of 11 000. During the period of the Dutch Republic, the status of Drenthe was unlike that of the other provinces. Although, in the course of the nineteenth century, Assen slowly tried to develop itself into a fully fledged regional capital, its economic structure around 1900 was still somewhat traditional, and it experienced almost no immigration.[8]

The other capitals shared a long tradition as regional central places, but there were important differences between them. Leeuwarden had to compete with other towns in Friesland, in particular Sneek, for dominance over the region. Groningen, on the other hand, had no competitors, and its economic influence even covered the north of the province of Drenthe. This further emphasizes the lesser position of Assen. Zwolle was surrounded by old cities (Kampen, Deventer, and to a lesser extent Zutphen), all of which attracted some central functions at the expense of Zwolle. Moreover, its influence was limited to the western part of the province of Overijssel because of the emergence of new industrial towns in the east. Arnhem faced some competition from the river city of Nijmegen, while 's-Hertogenbosch was already outnumbered by Tilburg (40 628 inhabitants). Maastricht is a separate case. Situated between Aachen and Liège, it industrialized rather early, with specialization in two branches in particular, pottery (Regout) and papermaking. Here we can see some indications of a dual economic structure.

In the other regional capitals, the regional functions seem to have been much more important than the functions within the urban network system. The concentration coefficients, the proportion of the labour force in a sector when compared to the proportion of the national population living in the city, generally do not exceed an average of 100.[9] Concentrations of 200 and over are seldom found in these cities, with the unsurprising exception of gas production, for gas factories were usually founded in this type of city. To complete the picture, in 1900 Leeuwarden had rather high concentration coefficients in printing (240), paper (658), trade (216), and the professions (209). Groningen was high in printing and the professions (285 and 229), Zwolle in chemistry (paint 265) and transport (241), Arnhem in printing (208) and insurance (235), 's-Hertogenbosch in printing (302), shoemaking (306) food and allied products (265), and Maastricht in pottery (2159) and paper (813).

Apart from Maastricht, the highest level of specialization was to be found in 's-Hertogenbosch. In Leeuwarden, the high concentration for paper was the result of just one factory.[10] With the exception of Maastricht, and perhaps 's-Hertogenbosch, there are no clear indications of the possible existence of a dual economic structure. This can only be confirmed, however, when the economic structure of the cities has been closely analysed, but such analyses are only available for two regional capitals, Leeuwarden and Groningen.[11]

Unfortunately, Leeuwarden is of no more than limited use for our analysis. Modernization there lagged far behind other regional capitals. In the long term it was to become the national centre for dairy industries, but around 1900 transport lines for milk were kept short by the lack of refrigeration techniques. The result was that dairy factories were scattered all over Friesland. In 1900, apart from the paper mill mentioned above, there were only two factories in Leeuwarden which had supra-regional importance: a strawboard factory (250 employees) and a shipyard annex machine factory (60 employees). One can hardly speak of a modern sector in Leeuwarden at that time,[12] and we therefore have to confine our analysis to the city of Groningen.

The economic structure and the labour market of Groningen

In Groningen specialization was rather effective, with four sectors achieving national importance during the second half of the nineteenth century.[13] First, there were firms in printing and publishing which specialized in the production of school books and almost monopolized this market in the Protestant parts of the country. Second, the food and allied products industry was well developed in Groningen. The main activity in the town's hinterland was arable farming: grains and sugar beet in the clay area to the north of the city, and potatoes in the peat area to the east of the city. As a result Groningen became the centre for the national grain trade. Except for a few flour mills, however, this did not stimulate any additional industry. The potato-flour industry was concentrated in the peat area itself, mainly in the small town of Veendam, but the headquarters of the largest producer of this product, W.A. Scholten, which also had factories in Prussia, Poland and Russia, was located in the city of Groningen. In 1896 this firm founded a large sugar beet factory in the adjacent village of Hoogkerk. In spite of an abundant supply of raw materials, specialization in the food industry was primarily in colonial products: a cane-sugar refinery (again a Scholten firm), cigars, tobacco, coffee and tea. The three factories producing canned meat were less exotic. Third, there was the production of bicycles. In the last quarter of the nineteenth century bicycles were produced in almost every city in The Netherlands, mostly by specialized smiths, but the industrial production of bicycles was concentrated in only a few places. The firm of Fongers in Groningen was one of these producers. The final area was the production of ready-made clothing. In the twentieth century Groningen became the main production centre for

Dutch ready-made clothing for men. Around 1900 some entrepreneurs, most of them Jews and/or German immigrants, started workshops, which in the long term would eliminate the domestic system.[14]

Specialization in Groningen did not generate the development of large enterprises. In 1910 there were only three factories with over 100 employees: the municipal gas factory, a hosiery factory and the Fongers bicycle factory. There were only ten factories, almost all belonging to the sectors mentioned above, which had between 50 and 100 employees. The average industrial firm did not exceed 20 employees. In municipal statistics the boundary between industrial firms and the crafts was fixed at 20 employees and/or the presence of mechanical drive, which meant that firms with just one employee could be defined as a factory. Alongside this, the statistics contain a list of the main craft firms. These are firms which, according to the view of the civil servant who compiled these statistics, had more than a local significance. Together these sectors contained about a quarter of the Groningen labour force in industry, the same proportion as in the preceding decades. We may therefore conclude that in the last quarter of the nineteenth century, the artisans constantly accounted for 75 per cent of the labour force in industry (10 082 in 1889 and 14 811 in 1910).

The proportion of mechanized firms, however, increased at the expense of the larger craft firms. This was caused by mechanization in traditional sectors and the founding of some new firms. Overall the number of firms having supra-local importance did not increase very much, but the number of employees did (Table 12.1).

The most important impetus for the change in proportion within the export sector was the introduction of gas motors in the 1880s and the electric motor in 1902 when the power station was opened. These kinds of energy were more suited than steam to the middle-sized Groningen industry. Printing companies and factories in the food and allied products sector adopted these especially early. The engines were primarily substitutes for unskilled human labour, but the electric motor also proved to be a stimulus to the creation of workshops in the clothing sector.

Statistics have enabled us to distinguish four sectors within the Groningen industrial economy: relatively large mechanized firms; small mechanized firms; larger craft firms; and artisans. We now have to ask whether any pattern of duality can be identified. The Amsterdam case would lead one to expect a division between the first category, relatively large mechanized firms, and the other three, and there are some signs supporting this view. Between 1895 and 1910, almost all of the 13 larger factories mentioned above built new factories on the outskirts of the city in which production was organized in a modern mechanized way that

Table 12.1 Mechanized and craft firms in Groningen

Trade	1870 M	1870 C	1890 M	1890 C	1910 M	1910 C
1 Bricks, pottery	2		2	11	3	8
Glass, mirrors		8		6		6
2 Printing			11	8	16	13
3 Candles, soap	1	5		4		1
Paint, oil	4	7	5	5	4	5
4 Wood: building	8		9		11	
Wood: furniture		34	2	28	3	27
5 Chemical cleaning			1		3	
6 Leather	1	12	1	2		1
7 Gold and silver		27		24	1	13
Copper, lead, tin		3		8	3	13
Iron (smiths)		96	2	52	1	53
8 Machinery/ships			1		1	1
Bicycles					1	16
Pianos		3				
9 Textiles (clothing)	2	45	4	15	3	9
10 Gas, electricity	1		1		3	
11 Vinegar		3	1	2		3
Salt		1	1			
Beer		3	1	2		
Bread			3		3	
Chocolate	1	1	3		4	
Chicory	3		3		3	
Gin	1	2	1	6	1	7
Groats, mustard	17		7		5	
Butter, margarine			2		6	
Lemonade		2		8	1	7
Flour	7		1			
Tobacco, cigars	4	42	26	14	16	21
Canned meat			1		3	
Cane sugar	1		1		1	
Total	53	294	90	195	96	204
Number of employees	555	1278	1342	712	2487	855

M = Mechanized industry; C = Crafts.
Source: Municipal Surveys.

included division of labour. In some branches there was a monopoly (sugar, canned meat), and in others the contrast with traditional predecessors was sharpened (bicycles, tobacco). Systematic research into wages in Groningen only covers the period up to 1874,[15] and by the 1870s wages in the few larger factories then in existence were already somewhat above the average. In the 1890s the Scholten sugar factory, in existence since 1862, paid the highest wages in the city. The wages in the clothing industry too were relatively high, with the replacement of the domestic industry structure which had embraced characteristics of the family economy. In the workshops one person had to earn a living for the whole family.[16]

In order to be certain that there was indeed a sharp division between the modern and the traditional sectors, it is necessary to take a closer look at the Groningen labour market. It is of particular interest to discover whether immigrants were less well off than the Groningen inhabitants, and in order to investigate this a number of samples from the Groningen databank were analysed. These were:

1. A random sample of Groningen heads of households for benchmark years (1870, 1880, 1890, 1900, 1910).
2. A random sample of immigrants, each covering the decade following the benchmark years mentioned above.
3. A cohort of Groningen-born children in 1880.

The main source for the construction of the samples was the Groningen Register of Population. Unfortunately, the indications of occupations in it are rather vague. For instance, almost no occupations are given for women and in many cases it is unclear at which firm people had their jobs. It is therefore impossible to determine exactly which people in the samples were employed in the 13 large firms. Nevertheless, the data do allow us to make a distinction between, on the one hand, a modern sector consisting of mechanized firms of different sizes and larger craft firms, and the traditional artisans on the other.[17]

The occupations of Groningen heads of household

In the previous section we concluded that in the period 1870-1910 75 per cent of the labour force in industry consisted of traditional artisans, taking both masters and men together. Bakers, butchers, shoemakers, tailors, carpenters, painters and blacksmiths were the largest categories. Other relatively frequent occupations were bleachers, watchmakers, plumbers, coopers and plasterers.

The period under review saw some significant changes. There was a relative decline in the number of masters. In 1870, of all the heads of households who held a job in industry, 41 per cent owned their own firm, whereas in 1910 the proportion was only 27 per cent. In the service sector, where the number of owners was much higher, a similar development took place - a fall from 72 per cent to 63 per cent. Changes were most striking after 1900. This change from employers to employees had already begun at the end of the eighteenth century. Although guilds were abolished during the Napoleonic years,[18] the greater part of the former guilds nevertheless tried to maintain their former prerogatives during the first half of the nineteenth century, especially the regulation of the labour market. They were not very successful, above all because members of the former guilds did not have a close connection with urban governments as had been the case during the *ancien régime*. The municipal council was now dominated by the urban elite, especially by people from the civil service, who lacked any real affinity with the crafts sector. Moreover, many craftsmen failed to fulfil their obligations. The former guilds were unable to cope with the new situation. Preoccupied with the local market, they did not have a keen eye for the new opportunities created by the formation of the urban network system. In this context it is not surprising that the greater part of the larger export industries was founded by people who came from outside the city.[19] Some craft masters did seize the new opportunities.[20] Certain pharmacists, for example, began the machine manufacture of paint, while some grocers established factories for coffee, tea and tobacco processing, and some smiths started the production of stoves and bicycles. Other masters tried to compete with their former guild brothers by hiring cheap labour.

As a result of these developments the crafts lost much of their attraction. First, the regulation of professional training disappeared. Second, the former guilds could not provide social security as they had done before because of the refusal of many former guild members to pay their contributions, and in any case most guilds had large debts. Finally, the intense competition in most sectors diminished the opportunities for employees to start a workshop of their own. Modern industry, however, offered many advantages: higher wages, funds for social security, and education in factory schools, while some firms even built houses for their employees. In 1870 35 per cent of Groningen male heads of households were artisans, but this percentage had fallen to 20 per cent by 1910. This decline coincided with a change in the composition of artisans. As Table 12.2 shows, some sectors, such as textiles and chemicals, disappeared completely, for the production of cotton and woollen fabrics, candles and soap had come to be concentrated in large factories elsewhere in the

Table 12.2 Composition of the Groningen group of artisans (masters and men) (%)

Occupation	1870	1910
Printers, bookbinders	4.6	3.9
Carpenters, bricklayers	25.9	20.5
Painters, plasterers	10.0	15.7
Soap and starchmakers, chemists	5.4	
Cabinetmakers	5.9	9.4
Coopers, coachbuilders	3.3	1.6
Brushmakers, basketmakers	3.3	
Tailors	4.2	11.0
Stonemasons, wood-carvers	1.7	0.8
Shoemakers, saddlers	8.4	6.3
Gold-, silversmiths	0.4	1.6
Plumbers, copper-, tin-founders, Blacksmiths	8.3	7.1
Weavers, bleachers, sail-, rope-makers	5.9	1.6
Bakers	6.7	11.0
Butchers	4.6	6.3
Others	1.3	3.1
N =	239	127

Source: Samples of Groningen heads of households in Kooij, *Groningen 1870-1914*, Appendix 1. Only men are included.

country. In fact, with labour productivity in the factories much higher, the crafts sector became both more traditional and more marginal.

As has already been observed, the proportion of industry in the Groningen occupational structure changed little, and the proportion of artisans within the industrial sector remained constant. The heads of household who left the crafts sector must therefore have been replaced by newcomers. Two questions therefore arise: which jobs attracted the attention of Groningen heads of households, and what role did immigrants play in the traditional sector? As far as the first question is concerned, in 1910 12 per cent of male heads of households worked in the modern part of industry, some of them as employers. Half of the employees had jobs with a relatively high status, for instance typographer or tailor's cutter, or were employed by the municipality at the gas factory, which was similarly considered to be a secure job. The others worked mostly in tobacco factories. This leads to the second question: did they also exclude immigrants from jobs in modern industry?

Immigrants and the Groningen occupational structure

In the last quarter of the nineteenth century, Groningen had a positive, but not very high, migration balance.[21] Most immigrants were young unmarried men and women who arrived from the Groningen countryside from which they were pushed out by the agrarian depression.[22] About a quarter of these immigrants did not obtain a regular job and many of them returned to their village of origin, while others migrated to other cities (Amsterdam). A smaller part stayed in the city and formed a labour reservoir. Which sector profited from this abundant supply of labour: the traditional, the modern or both? Table 12.3 gives the main occupations of the immigrants in the decade between 1900 and 1910. As it shows, most of the immigrants found a job in the services sector, almost all of the women becoming servants. The greater part of the married heads of household also found jobs there. A proportion of them belonged to the national circuit of civil servants. The national postal organization and the railway organization also transferred a number of their specialists to Groningen. It was industry more than trade which was the destination for young male immigrants, but they did not have easy access to the modern industrial sector. Most of them found employment in the traditional crafts sector. The bakers, the largest category within this group, mainly came to the city for the period of their apprenticeship and then returned to smaller places in the province of Groningen. The category of free labour consists mainly of navvies and others involved in preparing land for building, which was low-status work. They also did some work in the grain trade.

Table 12.3 supports the view that the locally born kept the better jobs for themselves. But they did not succeed in monopolizing the service sector, where better-educated immigrants had to be admitted. They seem to have been more successful in industry. In the traditional crafts sector the willingness of immigrants to accept low wages made them a real threat to the locally born. As a result competition increased and a number of small craft firms were unable to survive.

To establish whether the immigrants on the Groningen labour market did indeed function in a distinctive way, we now must consider the careers of those newcomers to the labour market who were born in the city itself.

Table 12.3 Occupations of Groningen immigrants 1900-10 (%)

Occupation	Men Modern	Traditional	Women Modern	Traditional
Agriculture		0.7		
Industry	4.4	25.3	0.8	0.8
Typographer	0.5			
Carpenter, bricklayer		2.5		
Architect	1.8			
Painter		2.1		
Chemist		0.7		
Cabinetmaker		0.9		
Brushmaker		0.2		
Woodcarver		0.2		
Clothing			0.8	0.8
Tailor		1.6		
Shoemaker		1.8		
Metalworker	0.2	2.4		
Shipbuilder	0.4			
Baker		9.1		
Butcher	0.2	0.9		
Brewer	0.2			
Watchmaker		0.7		
Cigar-maker	0.9			
Others	0.2	2.2		
Services	34.7		65.0	
Trade/ transport	15.9		3.0	
Hotel, catering	3.5			
Banking, insurance	0.9			
Professions	2.0			
Clerks etc.	1.3			
Domestic services	2.5		60.2	
Civil service[1]	5.3		1.5	
Religion	3.3		0.3	
Free labour	6.5		0.3	
No employment	28.4		32.9	
N =	552		337	

[1] Including teachers.

The Groningen birth cohort of 1880 and the labour market

There were 1577 children born in Groningen in 1880. A sample of one in two was taken, and the careers of these children were followed both in Groningen and elsewhere. Most children started work after they had completed primary school.[23] At the age of 30, most members of this birth cohort who were still alive had found their 'definite' destination. Most women had married and most men had found an occupation they would practise for years to come. Table 12.4 shows the occupations of the cohort members who stayed in Groningen.[24] Compared to the immigrants, the Groningen-born newcomers on the labour market took more jobs in modern industry, though a considerable percentage worked in the traditional crafts. Hardly any of them had jobs with the lowest status in building or as a baker or butcher. The largest group consisted of tailors, who at that time became part of the modern industrial sector. Moreover a number of artisans took over their parents' business, whereas the immigrants usually remained employees.

When compared to the immigrants, however, the cohort members were a little bit older. The age distribution of immigrants had two peaks. The median age of the single persons was 20-24 years, and of the heads of households about 40. The immigrants were thus at the beginning of their careers, though there are not many signs that there was much social mobility. It would perhaps be better, therefore, to compare the occupational structure of the cohort members with the occupational structure of the heads of households. Most of the cohort members had also reached that position by the age of 30.

Table 12.5 makes it clear that the modern industrial sector was dominated by heads of households and indigenous newcomers to the labour market. Immigrants could only add some specialism. In the services sector the differences between the three groups look smaller but a closer examination reveals some differentiation. The cohort members made up relatively the largest share of book-keepers, clerks, trade agents, travelling salesmen and so on, all jobs connected to modern industry. The largest proportion of immigrants gained low-status jobs in trade. The married immigrants, however, obtained specialized jobs in the civil service and at the railway company. In only one sector did unmarried immigrants successfully compete with the Groningen-born cohort members, and that was as shop-assistants in branch establishments of chain stores selling ready-made clothing. The women in the birth cohort also did remarkably well on the labour market. Most of them, of course, worked as domestic servants until marriage, but at the age of 30 there were still 35 women in the town with an occupation. Among these were nurses and teachers, and

Table 12.4 Occupations of male members of Groningen 1880 birth cohort at age 30 (%)

Occupation	In Groningen Modern	Traditional	Elsewhere Modern	Traditional
Agriculture		4.3		3.6
Industry	19.9	23.5	16.3	25.5
Printer, typographer	5.0			
Carpenter, bricklayer		2.9		
Architect	0.7			
Painter		7.1		
Cabinetmaker	1.4	0.7		
Woodcarver	1.4	0.7		
Clothing	0.7			
Tailor		7.9		
Silversmith		0.7		
Shoemaker		0.7		
Metalworker	3.6	1.4		
Butcher		1.4		
Tobacco/cigar-maker	4.3			
Gas producer	1.4			
Fireman	1.4			
Services	45.7		43.6	
Trade/transport	19.3 14.3			
Hotel, catering	2.1			
Insurance	0.7			
Professions	1.4			
Clerks etc.	2.9			
Domestic services	1.4			
Civil service[1]	3.6			
Free labour	6.4		10.9	
N =	141		55	

[1] Including teachers.

Table 12.5 Occupational structure of Groningen heads of households (1910), immigrants (1900-10), and birth cohort members (1910) (%)

	Heads of household		Immigrants		Birth cohort
Agriculture	1.5	(1.8)	0.7	(1.0)	4.3
Modern industry	12.0	(14.5)	4.4	(6.1)	19.9
Traditional crafts	20.0	(24.1)	25.3	(35.5)	23.6
Services	42.0	(50.6)	34.7	(48.4)	45.7
Free labour	7.5	(9.0)	6.5	(9.0)	6.4
Without employment[25]	17.0		28.4		

Notes: Only men included. Figures in parentheses indicate percentage excluding unemployed.

modern occupations such as photographer and telephone operator.

It therefore looks as if the Groningen-born cohort members were better off. They obtained relatively well paid jobs in industry and in the service sector, and even in agriculture where some took over their parents' market garden. There was no unemployment among this group, and the proportion of free labour was rather small. It was only a small proportion of this cohort, however, which was still in the town, for the greater part had left. We succeeded in tracing a number of the occupations which cohort members practised elsewhere, and the resulting picture is less optimistic (Table 12.4). Moreover, a substantial number of the cohort members (20 per cent) migrated directly to Amsterdam, while some arrived in that city at a later stage in their lives. Information about their occupations there is not yet available. Further research is needed to show whether in Amsterdam they ended up in the same situation from which they had tried to escape in Groningen.

Conclusion

The example of Groningen reveals some evidence of duality in regional capitals. Groningen, like most of the regional capitals, did not industrialize very much, but a modern sector consisting of some new specialist firms and of reorganized craft firms did emerge. As this essay has shown, jobs in this modern sector were for the most part filled by inhabitants of Groningen, most successfully by Groningen-born people.

They left some of the traditional craft occupations to the immigrants but nevertheless kept a large share in this traditional sector as well. Alongside the dualism in the 'industrial' labour market, the services sector showed a three-way division. The lowest-status jobs in domestic service and trade were mainly reserved for single immigrants, while the other jobs were very popular among the Groningen heads of households. High-status vacancies at the university, in the civil service and the professions, however, were increasingly filled by immigrants.

As a result of this development of duality within the industrial labour market, the status of artisans fell dramatically. The craft sector was unable to compete with the higher profits and wages in the industrial sector and its dynamic of innovation, better career prospects and social security. It is no wonder that many artisans tried to find work in this industrial sector, whether as employer or employee. If only a few of them succeeded, their efforts to do so emphasized the changing position of artisans in society: from well educated, politically influential regulators of labour conditions with a well developed social safety net, into an obsolete group of rather old-fashioned producers.

Notes

1. J.A. de Jonge, *De industrialisatie in Nederland tussen 1850 en 1914*, Amsterdam: Scheltema & Holkema (1968).
2. J.A. de Jonge, 'The role of the outer provinces in the process of Dutch economic growth in the nineteenth century', in J.S. Bromley and E.H. Kossmann (eds), *Britain and the Netherlands: IV: Metropolis, Dominion and Province*, The Hague: Martinus Nijhoff (1971), pp. 208-26.
3. J.L. van Zanden, *De industrialisatie in Amsterdam 1825-1914*, Bergen: Octavo (1987).
4. Ad Knotter, *Economische transformatie en stedelijke arbeidsmarkt. Amsterdam in de tweede helft van de negentiende eeuw*, Zwolle: Waanders (1991).
5. Henk van Dijk, *Rotterdam 1810-1880. Aspecten van een stedelijke samenleving*, Schiedam: Interbook International (1976).
6. For an analysis of the role of the infrastructure in the integration of one region into a larger framework, see Marcel Clement, *Transport en economische ontwikkeling. Analyse van de modernisering van het transportsysteem in de provincie Groningen (1800-1914)*, Groningen: Wolters Noordhoff (1994). An English summary of his dissertation can be found in Marcel Clement, 'Transportation Networks in the Dutch Province of Groningen', *Economic and Social History in the Netherlands*, 5, 1993, pp. 103-29.
7. P. Kooij, 'Peripheral cities and their regions in the Dutch urban system until 1900', *Journal of Economic History*, 48, 1988, pp. 357-71.
8. Mieke Lems, 'Migratie in Assen', master's thesis, University of Groningen, 1994.

9. For a full account of concentration numbers, see Kooij, 'Peripheral cities', and P. Kooij, 'Urbanization. What's in a name?', in H. Schmal (ed.), *Patterns of European urbanisation since 1500*, London: Croom Helm (1981), pp. 31-61.
10. Rolf van der Woude, *Leeuwarden 1850-1914. De modernisering van een provinciehoofdstad*, Leeuwarden: Fryske Akademy (1994).
11. Ibid.; P. Kooij, *Groningen 1870-1914. Sociale verandering en economische ontwikkeling in een regionaal centrum*, Assen/Maastricht: Van Gorcum (1987).
12. As a result Leeuwarden had a relatively large migration deficit. Many inhabitants migrated to Amsterdam where for the greater part they found jobs in the traditional part of the dual labour market.
13. Pim Kooij, 'Groningen: central place and peripheral city', in Pim Kooij and Piet Pellenbarg (eds), *Regional Capitals. Past, Present, Prospects*, Assen: Van Gorcum (1994), pp. 37-63.
14. Specialization in printing and clothing is shown to some extent by the Groningen concentration coefficients (285 and 189 in 1900). Specialization in bicycles, however, is not shown by the coefficients since in the statistics this branch was a part of the much larger metal producing sector (98). The same was true for the specialist trades within the food and allied products sector (143).
15. Albert Beetsma, Daan van der Haer and Herman de Jong, ' "De bron van alle zorg; de zorg van alle dag". Een onderzoek naar de lonenstructuur in de provincie Groningen, 1854-1874', master's thesis, University of Groningen, 1983.
16. See A.A.P.O. Janssens, *Family and social change. The household as a process in an industrializing community*, Cambridge: Cambridge University Press (1993).
17. Knotter was also faced by the problem of insufficient detail on occupations. For the much larger city of Amsterdam, however, it was possible to focus on sectors which were almost completely modern and others which remained traditional. Mixed sectors do not play a major role in his analysis.
18. This took place in 1798 and was confirmed by King William I in 1818.
19. Kooij, *Groningen*, p. 338.
20. Here lies a parallel with Hungary; see the contribution by Vera Bácskai to this collection (Chapter 10).
21. Kooij, *Groningen*, ch. 3, pp. 82-190, and Kooij 'Groningen', pp. 51-2.
22. In the decade 1900-10 76 per cent of the immigration units consisted of single persons, with an equal proportion of men and women. Twenty-four per cent consisted of heads of households with their family.
23. Pim Kooij and Jules L. Peschar, ' "Tot maatschappelijke en christelijke deugden". Een longitudinaal onderzoek naar schoolkeuzen in de Groninger geboortencohort 1880', in Pearl Dykstra, Pim Kooij and Jan Rupp, *Onderwijs in de tijd. Ontwikkelingen in onderwijsdeelname en nationale curricula*, Houten/Zaventem: Unieboek (1992), pp. 90-110.
24. The sample consists of 419 men and 369 women. After 30 years 152 of the men had died (35 per cent); 146 (35 per cent) still lived in the city; the others (30 per cent) had migrated (and sometimes died). Among the women 27 per cent had died, 37 per cent still lived in the city, and 36 per cent lived elsewhere (or had died there).

25. In every benchmark year about 17 per cent of the heads of households in 'normal' years, and 25 per cent during the agrarian depression, did not have an occupation. The greater part of them however were not looking for any job, as most of them had retired. Yet a number of heads of households belonged to the 'reserve army'. Some of them were immigrants and had married in the city. As a result the proportion of what was called 'free labour' was rather high.

Index

Aachen 51
Albistur, M. 151
Amsterdam 239, 244, 253
Anglomania 100, 103-4
apprenticeship 7, 65-7, 78, 121, 125, 157, 174, 176, 198, 249
  geographical origins, apprentices' 184-6
  women 67, 74, 154, 160
Archer, Ian 82
aristocracy
  consumption patterns 93-4, 99-107, 112-13
  debts 109-12
  and rural industry 201-2
  sociability 107, 109
  see also court society; hôtel, aristocratic
Armogathe, D. 151
Artéus, Gunnar 140
articles de Paris makers 220-21, 226, 228-35 *passim*
artificial flower makers 28, 221, 228-9, 230
artisanal production 8-9, 26-32, 161-2, 167-8, 200, 203-5, 220-21, 223-6, 239-43, 247-8
  artificial power 240, 244, 245
  conditions 206-9
  division of labour and specialization 30-32, 96, 101, 104-5, 108, 162, 228-30
  fabrique 27, 221, 226, 227-32
  flexibility 28-9, 30, 227, 232
  and industrialization 26-32, 63-4, 200, 215-16, 239-40, 241-3
  and merchants 22, 27-8, 80, 161-2, 202, 224-6
  see also artisanal production, *fabrique*; guilds, merchants' role
  numbers of enterprises 26, 206, 223-8 *passim*, 237
  relative decline 4, 31-2, 206, 208-9, 212, 226, 231, 247-8, 249

rural production, *see* artisans, rural
sub-contracting 8-9, 21, 27-8, 29, 30-32, 77, 165
see also dual economy; sweating
artisans
  career 7-9, 13, 66, 121-2, 186
  see also journeymen, access to mastership
  credit and debt 10, 27, 109-12, 209-10
  definitions and meanings 4-15, 56-8
  distribution in towns 95-9, 114, 120, 128, 233-6
  free 117
  geographical mobility 172-5, 179-94, 248
  see also journeymen, geographical origins; journeymen, travel
  honour 10, 11, 15, 60-62, 69
  identity 2, 11-13, 42, 49, 58-60, 62, 144, 203
  income 206-9, 246
  pictures of 102, 105, 108-9
  poverty 125
  rights 88-9
  rural 16, 42, 52, 181, 200-203, 206, 207
  sources for study of 5, 7, 9, 13, 16, 93, 94-5
  work on customers' premises 9, 107-9
  see also artisanal production; craft culture; master artisans; ritual, artisanal; skill
Arts and Crafts movement 33
Augsburg 3, 14, 17, 49-50

Babeau, Albert 1
bakers 124, 140, 153, 157, 182, 204, 246-52 *passim*
bankruptcy 110, 111
Basel 12
basketmakers 81, 82-3, 248

INDEX

Bayeux 11
Bayonne 117, 123
Berg, Maxime 28
Berlin 30, 184
bicycle manufacturing 31, 243-7
    passim
Birmingham 28
Blanqui, Adolphe 30
Bochum 26
Boissieu, Comte de 2
Bologna 16
Booth, Charles 31
Bordeaux 19, 21, 116-26
Bossenga, Gail 161
Bourdieu, Pierre 59
Brants, Victor 28
Braverman, Harry 64
bricklayers, see building trades
Buda, see Budapest
Budapest 184, 205-14 passim
building trades 9, 31, 67, 95-103
    passim, 113, 163, 164, 203,
    204, 208, 211, 216, 240,
    246-52 passim
    see also joiners
burgher estate and status 12, 45,
    132-8, 146-7
burgher rights, see civic rights
burghers' guard (Malmö) 140
Burke, Peter 56
butchers 120-21, 128, 153-9
    passim, 164, 177, 182, 204,
    208, 246-52 passim

cabinet makers, see furniture trades
Caen 10
canuts, see silk-weavers
Caron, François 26
carpenters, see building trades
ceremony, see symbolic order
Cerutti, Simona 6, 21, 57, 122
chambrelan, see guilds, illegal
    production
Chaussinand-Nogaret, Guy 94
Chevalier, Bernard 60
Chevalier, Michel 30
chimney-sweeps 184, 185, 208
civic rights 17, 48-9, 136, 177, 201,
    213-14
civic ritual 17-18, 32, 149
Clark, Alice 151

clockmakers 82, 83
clothing trades 30, 95, 97, 98, 108,
    113, 155, 160-67 passim, 212
    214, 229-34 passim, 240,
    243-4, 250, 252, 255
    see also tailors
coachbuilders 103-6
Cochin, Augustin 229
Collins, J.B. 151, 153, 154
Cologne 13, 24, 47
compagnonnages, see journeymen's
    associations
coopers 117, 121, 182, 246, 248
corporate ideals 23-5, 60-61, 65
corporations, see guilds
Cottereau, Alain 30
court society 92-3
craft culture 11-12
craft histories 1-2
craft museums 2
crown 141
    artisans appeal to 43-4
    visit to Malmö 139-41
cultural analysis 58-60, 69-70
cutlers 8

Darnton, Robert 62
Davis, Natalie Zemon 153
De Jonge, J.A. 239
De Vries, Jan 163
Debrecen 207, 208
decorating trades 99-103 passim,
    108, 113
    see also building trades
Deventer 18, 242
Dijon 8, 12, 65-6, 151
dressmakers, see seamstresses and
    dressmakers
Du Maroussem, Paul 31
dual economy 31, 239-41, 244-54
Düsseldorf 24

Edinburgh 28
Estates General (France) 123
Esztergom 207, 208

Fairchilds, Cissie 161
family 7, 10-11, 165, 166, 176, 211
    see also household
Farge, Arlette 9, 62, 66
Farr, James 151

INDEX 259

faux ouvriers, see guilds, illegal production
Ferry, Jules 220
food trades 97, 98, 113, 152–67 *passim*, 177, 242–7 *passim*, 250, 252
Frankfurt am Main 51–2, 62
French Revolution 19–20, 116, 123–6
Funck-Brentano, Théophile 18
funerals, *see* religion, funerals
furniture trades 24, 30, 33, 81, 96, 98, 100, 162, 174, 175–6, 181–3, 185, 187–8, 204, 206, 225, 245–52 *passim*

Gaillard, Jeanne 228–9
Gallinato, B. 121
Garden, Maurice 8, 27, 167
gardens, aristocratic 100, 101–2
geographical mobility, *see* artisans, geographical mobility; journeymen, travel
glovemakers 136–7, 144, 204
golden age, artisanal 1–4
goldsmiths 66, 82, 119, 128, 154, 157, 162, 174, 182, 245, 248
Gothenburg 132, 134, 135
Great Exhibition (London, 1851) 30
Great Fire (London, 1666) 78
Groningen 31, 239 54
guilds 10, 12–13, 18–25, 57, 60–63, 65–9, 72, 96, 101, 108, 201–2
 admission 66, 119–20
 in Austria 19, 20, 22, 175–9
 in Belgium 19
 in Britain 19, 20
  see also livery companies
 control over production 21–2, 42, 201–2, 206, 212
  see also guilds, inspection
 and countryside 202, 203
 decline and abolition 4, 10, 19–24, 123–6, 167, 212–13
 divergence from ideals 21–2, 121–2, 126
 and family 10–11
 in France 19–20, 22, 116–26 *passim*, 51–68 *passim*
 in Germany 19, 20, 22, 24, 41–53
 in Hungary 18, 20, 22, 201–3

 idealizations of 2, 16, 18, 212
 illegal production 21–2, 67, 80, 118–19, 162–3, 165–6, 167, 173, 181–2
 inspection 75–6, 78, 79–85, 86–8, 118
 and labour discipline 63, 65, 66, 164
 masterpiece 119, 128
 merchants' role 22, 79, 83, 161–2, 178
 mixed trade 12, 203
 numbers of 12, 117, 119, 122, 177, 196
 rights of daughters 153–5, 159–60, 169
 rights of sons and sons-in-law 42, 66, 128, 153–5, 181
 size 12, 177, 196
 state policy towards 12, 119–20, 177–8, 201
  see also laws
 in Sweden 19, 20, 22, 24, 133, 144
  see also St Canute's Guild (Malmö)
 and towns 16–18, 20
 unofficial guilds 202
 women's guilds 13–14, 153
 women's role in 13–14, 68, 151–68 *passim*
  see also widows; women
 see also corporate ideals; French Revolution; municipal government, and guilds; privileged places; processions
gun makers 28
Györ 207–210 *passim*

haberdashers 95
Hannover 47
hatmakers 62, 65–6
Haussmann, baron Georges 218–22 *passim*, 226
Haussmannization 29–30, 218–22, 235–6
Hegel, Georg Wilhelm 67
Henson, Gravener 88
's-Hertogenbosch 241, 242
Hirsch, Jean-Pierre 20

INDEX

Hobsbawm, Eric 84
home towns 15-16, 47
honour, see artisans, honour
horse-related trades 95, 97, 98,
    103-6, 108, 113, 158, 162
hôtel, aristocratic 92-113 passim
household 9-11, 68-9, 163
    as ideal 3, 9, 14
    employees live in 7, 10, 121, 165
    see also family; journeymen;
        patriarchy
Hufton, Olwen 166
Husson, François 2

illegal production, see guilds, illegal
    production
industrialization, see artisanal
    production, and
    industrialization
industry, larger-scale 161, 162,
    22-7, 230-32, 239, 242-8
    passim
instrument makers 83

jewellery trades 110, 220, 230, 232,
    234
joiners 96, 108-9, 124, 154, 157,
    164, 165, 167
journeymen
    access to mastership 8, 21, 121-2,
        124-5, 169, 203, 249
    drinking 7, 62-3, 217
    geographical origins 186-8,
        249-53
    hiring 129, 158
    marriage by 8, 125, 129-30, 179,
        197
    ratio to masters 175-6, 203-4,
        247
    travel 3, 7-8, 49, 173, 186-91
        regulation 8, 178-9, 188-9,
            192-4
        rituals 7, 63, 189-91, 194
    see also household, employees
        live in; journeymen's
        associations; masculinity;
        workplace relations
journeymen's associations 21, 25,
    37, 60-63, 64, 88, 124, 126,
    127
Joyce, Patrick 58

Kaplan, Steven 65
Knotter, Ad 240

labour market 15, 61, 63, 172-4,
    181-2, 190-91, 240, 246-54
laundresses 98
laws
    General Handicrafts Act (Holy
        Roman Empire, 1731) 178,
        192-3
    loi d'Allarde (France, 1791) 19,
        24, 116, 124
    loi Le Chapelier (France, 1791)
        24, 116, 124
    Swedish national guild code
        (1720) 133
    Statute of Artificers and
        Apprentices (England, 1563) 23
    Turgot's Edict on corporations
        (France, 1776) 10, 20, 116,
        119, 151, 167
leather trades 78, 152, 157, 206,
    211, 225, 245
Le Playists 2-3
Leeuwarden 241, 242-3
Lille 20, 22, 161
livery companies, London 75-89
    yeomanry 77, 80, 84, 86
    see also guilds, in Britain; guilds,
        inspection
local government, see municipal
    government
locksmiths, see metal trades
London 21, 27, 30, 31, 75-89,
    103, 184
Lübeck 12, 42, 52
Luddism, see machine-breaking
luxury trades 97, 98, 113, 162, 183
Lyon 21, 22, 27, 153, 161, 167, 184

machine-breaking 23, 84-9
Malmö 12, 17, 132, 136-47
Marcelin, Paul 31
Marglin, Stephen 64
marriage 9-10, 130, 154-8 passim,
    169, 184
    see also journeymen, marriage by
Martin, Gaston 162
Marx, Karl 58
masculinity 13, 15
    see also women, gender division

of labour
master artisans
  associations 24
    see also guilds
  legal status 23-4
    see also guilds
  numbers of 117, 125, 137, 175-6, 177, 181, 206, 247
  wealth inequalities amongst 120-21, 125-6, 161-2, 205-12
    see also artisanal production; artisans; women, as masters; workplace relations
Mayhew, Henry 31
Ménétra, Jacques-Louis 3, 4, 5, 11, 62, 164
merchants 133-6, 140, 142-5, 204-8 passim
  see also artisanal production, and merchants; guilds, merchants role
Mercier, Louis Sébastien 3, 108
metal trades 8, 11, 12, 27-31 passim, 42, 78, 82, 98, 99-103 passim, 104, 108, 119, 124, 152-64 passim, 182, 200-208 passim, 225-31 passim, 235, 245-52 passim
migration, see artisans, geographical mobility
milliners 13, 14
Montesquieu, Charles de Secondat, baron de La Brède et de 103
Morris, William 33
municipal government 20, 247
  artisanal role in 213-15, 217
  ceremonies 139
  and guilds 15-18, 20, 41-4, 46-7, 118-19, 123-4, 127, 176, 201, 213, 247
  in Sweden 133-8, 148-9
  see also civic rights; municipal politics; processions
municipal politics 41-53, 215
Musgrave, Elizabeth 67

Nantes 14, 21, 152-68
neighbourhood 7, 9, 220-36 passim
Nîmes 31
Norrköping 132

Norwich 12
Nottingham 88-9

occupational titles 5-6, 65, 73-4
occupations, multiple 6-7, 11-12
Orientalism 100
Orleans 22, 161
outwork 22, 27-8, 30-31, 77
  see also artisanal production, sub-contracting

Paris 3, 21, 27-8, 29-30, 31, 92-113, 218-36
  see also Haussmannization
patriarchy 9, 154, 212
Perdiguier, Agricol 5
Perrot, Jean-Claude 10
Pest, see Budapest
Pétain, maréchal Philippe 3
petitions 43, 50
Pied, E. 160
pluriactivity, see occupations, multiple
politics
  artisanal 24-5, 41-53
  definition 45-6
  see also municipal politics
printers 8, 62, 174, 242-55 passim
privileged places 117-18
processions 146
  guild 2, 146, 150, 156
  municipal 32, 139
  see also crown, visit to Malmö
property
  mastership as 124
  ownership by artisans 210
  see also skill, as property
Pytteroen, Oscar 18

Quataert, Jean 68-9

railways 241
Rancière, Jacques 5
Reddy, William 57, 59
Reid, Donald 57
religion 7, 32
  funerals 142-5, 146
  pew renting 142, 143, 150
Rennes 155
retailers 107, 109, 227-8, 232, 238
Reval (Tallinn) 41-5, 47, 48

ribbon-weavers, *see* silk-weavers
Ricardo, David 58
riots and protest 43, 48, 51-2, 84-6
ritual, artisanal, 62-3
  see also journeymen, travel
ritual, civic, *see* civic ritual
Roche, Faniel 3, 98, 110
Roeck, Bernd 49
Roper, Lyndal 3
Rousiers, Paul de 11
Rule, John 64

Sabel, Charles 29
saddlers, *see* horse-related trades
Sahlins, Marshall 58
Sahlins, Peter 57
Saint-Etienne 28
Schilling, Heinz 51
Scott, Joan Wallach 57
seamstresses and dressmakers 13, 14, 64, 118, 126, 127, 166
searches, *see* guilds, inspection
secondhand-goods dealers 156
Sewell, William 25, 57
Sheffield 8, 28
shoemakers 5, 16, 24, 27, 29, 30, 43, 49, 118, 120, 123, 128, 154, 164, 165, 174, 181-3, 201, 242, 248, 250, 252
silk-weavers 27, 28, 84, 88, 161, 183-4, 187-8
Sitte, Camillo 2
skill
  as cultural construct 63-9
  and gender 13, 64, 68-9, 158
  as property 23, 64
Smith, Adam 58
Social Catholics 2-3
Solingen 28
Sonenscher, Michael 8, 27, 57, 64, 161, 162, 165
Sopron 209
specialization, *see* artisanal production
spectacle makers 81, 83
St Canute's Guild (Malmö) 139, 141, 144, 147
state, *see* guilds, state policy towards
steam power 26, 39

Stockholm 42, 43, 132, 134, 135
stocking-knitters 87, 161, 189-90, 191, 193-4, 198
Stopp, Klaus 180
Strasbourg 12
strikes 21
  see also riots and protest
sweating 30-32, 226
symbolic order 131-2, 138-47
Szeged 207-8

tailors 5, 11, 24, 27-8, 29, 42, 98, 118-26 *passim*, 128, 155-60 *passim*, 165, 166, 174, 181-3, 201, 204, 205, 246-52 *passim*
taxation 123, 135, 177, 206
  patente 124, 223-38 *passim*
terminology 25, 56, 60-61, 124
textile trades 11-14 *passim*, 22, 27, 78, 82, 84-8, 152-67 *passim*, 173, 175, 177, 182, 183, 205, 225, 229, 239, 245-50
  see also silk-weavers
Thamer, Jans-Ulrich 62
tools 62
Toulouse 28
Tour de France, *see* journeymen, travel
towns 15-18, 132-3
  economy 29-30, 132, 160-61
  regional capitals 239-43
  social structure 137
  see also civic ritual; guilds, and towns; *Haussmannization*; home towns; municipal government; municipal politics
tramping, *see* journeymen, travel
Truant, Cynthia 62, 153, 167
Turgot, Jacques 10, 20, 116, 119, 151
Turin 6-7, 11, 18, 20-21, 122, 149

Van Dijk, Henk 240
Van Zanden, J.L. 239
Versailles 94
Vienna 2, 20, 172-94

Walker, Mack 15-16, 47
Wanderjahre, *see* journeymen, travel
white-collar workers 3, 250, 252
widows 14-15, 118, 152-7 *passim*,

164, 169, 170-71
Wiesner, Merry 15, 62, 68
wigmakers 119-26 *passim*, 130, 160
women 13-15, 86, 142, 150, 151-68 *passim*
  in artisanal enterprise 13, 151, 160, 163
  as dishonourable 15, 61-2, 69
  in labour force 15, 163
  legal status 13-14
  as masters 13-14, 153-60 *passim*, 164
  wives in business 2, 9-10, 14-15, 155, 156, 163-4, 166-7
  see also apprenticeship; family; guilds, women's role in; guilds, rights of daughters; marriage; skill: widows
workplace relations 3-4, 9, 21, 121, 164
  idealizations of 2-3, 26
  see also strikes

yeomanry, *see* livery companies, London
York 6

Zarca, Bernard 4
Zeitlin, Jonathan 29